Modelling of GPCRs

Andrea Strasser • Hans-Joachim Wittmann

Modelling of GPCRs

A Practical Handbook

Springer

Andrea Strasser
University of Regensburg
Institute of Pharmacy
Dept. of Pharm./Med. Chemistry
Regensburg
Germany

Hans-Joachim Wittmann
University of Regensburg
Faculty of Chemistry and Pharmacy
Regensburg
Germany

ISBN 978-94-007-4595-7 ISBN 978-94-007-4596-4 (eBook)
DOI 10.1007/978-94-007-4596-4
Springer Dordrecht Heidelberg London New York

Library of Congress Control Number: 2012938366

Printed on acid-free paper

Springer is part of Springer Science+Business Media (www.springer.com)

Preface

Biological cells function as elementary building blocks for living individuals. All compounds, essential for establishing and maintaining life processes are to be produced inside the cells. This makes it necessary for molecules and ions to pass the cell membrane in order to take part or to support the appropriate biochemical reactions. Furthermore, a lot of regulatory processes control the complicated sequences of the molecular reaction cycles and signal cascades and may be influenced by information in form of physical effects or chemical compounds coming from the environment of the individuals. So, all what we call life is subject to biochemical processes and may be described by thermodynamic and kinetic concepts. Energetic and entropic aspects were therefore used in a larger extent to explore the behaviour of chemical compounds addressing G protein-coupled receptors residing in the cell membrane. In this context, drug design in the past was done by chemical synthesis and pharmacological testing afterwards, hoping to obtain a powerful new active compound. But in order to have specific drugs, exhibiting a minimum of side effects and to reduce costs and time of research and production, a deeper insight onto processes linked with the interaction of ligands and receptors on molecular level is necessary. So, nowadays a scientist working on the field of drug design has to use the physicochemical concepts to successfully predict the properties of compounds. But an increasing knowledge of the processes determining the behaviour of the interaction between ligands and receptors reveal a great complexity of this research field. Computational methods have to be used in order to describe quantitatively the processes setting up the network of ligand-receptor-interaction and the related signal cascades. Working on the field of GPCRs, theoretical concepts have to be developed and a large number of related programs have to be designed and it turns out that the operation system UNIX/LINUX is the best solution to do all this work in a highly efficient manner. Thus, we got the idea to present not only a review of methods and results concerning the modelling of GPCRs, but to establish a practical guide for researchers interested in this field. Realizing the great importance of the work in computing, we included a chapter designed as an overview of the most important UNIX/LINUX commands and present a lot of solutions concerning computational problems. We hope, researchers will comprehend the benefit of the operating system. All commands and scripts presented in this book were developed very carefully. Nevertheless we do not give any warranty for correctness.

Contents

Chapter 1
Introduction

The knowledge about conformation of proteins and distinct interactions between a ligand and it's target protein is necessary to explain pharmacological data on a molecular level. Additionally, based on this knowledge, it may be possible to develop new, potent drugs more efficiently. But how get these insights on a molecular level? Several experimental techniques, like mutagenesis studies combined with pharmacological investigations may give hints about amino acids, being important for stability of a protein or being important for the interaction between ligand and protein. But these studies exhibit no information about energetics and hydrogen-bond-networking for example. Other techniques, like determination of structures of proteins or protein-ligand-complexes by NMR or crystallography are very useful to obtain information about secondary, tertiary or quartary structures of proteins (http://www.pdb.org). However, these experiments are time-expensive and cannot be performed for each system of interest like on an assembly line. Additionally it has to be taken into account, that crystal structures represent a solid phase, but proteins are in general in solution and exhibit a kind of dynamical behaviour. This is taken into account by molecular dynamic simulations of a protein in its natural surrounding. Additionally, with several distinct molecular modelling techniques, ligand-receptor interactions for example, can be simulated in a reasonable time and insights onto interactions on molecular level can be obtained. Furthermore, some techniques, like 3D-QSAR (Brown et al. 2006; Dudek AZ et al. 2006; Gedeck et al. 2008; Scior T et al. 2009) allows predicting affinities also in context with GPCRs (Strasser et al. 2010a; Silva et al. 2011). However, molecular modelling results should be, when ever possible, compared with experimental data in order to judge predictive quality. To combine the experimental results with computational methods in order to understand and moreover to predict the behaviour of systems involving chemical reactions, it is necessary to establish a link between macroscopic quantities, like equilibrium and rate constants, thermodynamic quantities like ΔH^o and ΔS^o, which are available from experimental methods (Leavitt S et al. 2001; Wittmann et al. 2009; Torres et al. 2010) and microscopic properties, like energy levels, which result from the interactions of the nuclei and electrons comprising the distinct particles of the system of interest. This task is not a simple one especially when certain properties for example of a ligand interacting with the receptor are to be depicted as thermodynamical

A. Strasser, H.-J. Wittmann, *Modelling of GPCRs*,

DOI 10.1007/978-94-007-4596-4_1,

quantities. So, there were a lot of efforts in the past to classify ligands as agonists or antagonists with the help of ΔH^o and ΔS^o. One attempt to distinguish between the two groups of ligands is based on the term enthalpy or entropy driven association process. Enthalpy driven means $\Delta H^o < 0$ and $\Delta S^o < 0$, entropy driven is indicated by $\Delta H^o > 0$ and $\Delta S^o > 0$, whereas $\Delta H^o < 0$ and $\Delta S^o > 0$ is called enthalpy-entropy driven (Weiland et al. 1979; Wittmann et al. 2009). But by investigating the extensive data material no definite discrimination between agonists and antagonists is possible on this basis. The crucial point results from the fact that ΔH^o and ΔS^o determine the affinity of a ligand investigated in a binding assay. But if we talk about agonists or antagonists, we put our focus on the efficacy, which will be determined from corresponding assays. To combine binding properties like ΔH^o or ΔS^o with quantities, describing the efficacy will not lead to satisfactory results. Thus, is there a chance at all to predict the binding behaviour of a ligand on the base of the thermodynamical concept, discussed in Chap. 10? As a first step, we have to establish a binding model based on our knowledge or intuition of the interaction between the ligand and the receptor. X-ray based structures are the best choice up to now to get structures of the interesting biochemical system, which would then be utilized to calculate ΔH^o and ΔS^o or rate constants for comparison with the experimentally determined values and to validate a particular model. Making use of statistical mechanical concepts (see Chap. 7.1), the central quantity is the potential energy of the system from which we are able to calculate the phase integral and thereafter the chemical potential, which governs the chemical behaviour of an arbitrary species. These concepts are adopted in the framework of the quantum mechanical concept by calculating the so-called partition sum (see Chap. 7.1). Here, we also have to define the potential energy of the system of interest and then we have to solve the corresponding Schrödinger equation to get the allowed energy levels. But up to now, it is impossible for such large systems, comprised of ligand, receptor, membrane, water and ions to do such ab initio calculations in an acceptable time. To simplify the calculation procedure, a stable state is defined as a energy minimum of the so-called potential energy surface, represented by the potential energy as a function of all coordinates of the particles present in the system. Starting from a first guess of a structure, minimizing the potential energy with respect to the coordinates, will lead to a final structure from which we are able to derive a set of properties. Even this modified procedure leads to a very time consuming calculation. Thus, ab initio methods are not suitable to handle biochemical systems. However, sometimes, this method is used in context with GPCR research (Carloni et al. 2002; Mehler et al. 2006; Jongejan et al. 2008). The so-called semiempirical methods use potential functions based on some experimental insight to find local minima across the potential energy surface (Stewart 1989; Stewart 2004; Lipkowitz et al. 2007). This concept reduces the computational time but introduces a new problem based on the choice of the semiempirical method, which seriously influences the computed results. In order to get a very simple functional form of the potential energy resulting in small computational times, molecular dynamics (molecular mechanics) makes use of so-called force fields (see Chap. 9), which entirely depend on empirical quantities, so the quality of the results strongly depends on the experimental parameters used to define the particular force field. To combine the well founded theoretical concept of quantum mechanics with the advantage of a short computational time, hybrid methods, such as quantum mechanics/molecular mechanics (QM/MM) concept are introduced (Monard et al. 1999). The interesting part of the system is calculated using the principles of quantum mechanics, whereas the remainder of the system is treated by the methods of molecular mechanics. To take advantage of this method we have first to define the boundary between the "quantum mechanical region" and "classical region" and secondly, we have to establish a connection between the two regions, which is done by introducing so-called link atoms. A further improvement of the hybrid methods is developed in the framework of the moving domain quantum mechanics/molecular mechanics (MoD-QMMM). To gain deeper insight into the basics of the hybrid methods, the reader is referred to the literature (Gascon et al. 2006; Menikarachchi et al. 2008). Searching the potential energy surface for minima, any of the mentioned methods will find only local minima. To identify the most stable configuration of the system of interest, the global minimum of the potential energy surface should be detected, but up to now, no reliable algorithm, solving this problem is available. Thus, to get enough information about the system of interest, multiple scans have to be done from distinct starting structures. But in this context, the question arises, whether these different configurations have to be linked by equilibrium processes or not. Doing so, we will get a very large set of structures from which we have to explain the interaction between the ligand and the receptor. A further crucial problem appears, when the entropic contributions are to be evaluated. Molecular mechanics methods totally lack the calculation of such terms, whereas quantum mechanical based methods allow for estimating the entropy term of a system in principle, which is given mainly by the vibration modes. So, if we deal with a system comprised of N sites we have to determine $3 \cdot N - 6$ vibration modes, i.e. for N = 10,000 there are nearly 30,000 vibrational terms to be computed. Further on, there is another problem arising from the modes belonging to transition states. Since we are interested in equilibrium states and get a lot of transition modes, we have to change the geometry of our system in a way that only real vibrational modes appear, which is a very tedious task. Many of the vibrational modes describe internal rotations around bonds, characterized by low frequencies and therefore make an unacceptable large contribution to the overall vibration energy. An exact treatment of this motion is not available up to now. The prediction of the entropy term ΔS^o in this context is a very difficult matter and consequently the results are not reliable. Because of this difficulties, in almost all studies, based on the mentioned methods, also called single point calculations, only the potential energy terms or the allowed energy levels of the system are used for a qualitative discussion of its behaviour. To overcome the problems caused by single point calculations, molecular dynamic studies (MD) on biological systems have to be carried out. These methods make use of the equation of motion, introduced by Newton, to compute the time evolution of a system. For calculating thermodynamical quantities, the reader is referred to Chaps. 7 and 10. Up to now, processes with time constants in the range of some μs are subject to MD simulations, so processes taking place with time constants in the range of ms or larger, like diffusion processes in solutions, cannot be captured by this method. Furthermore, force field methods are unable to handle processes accompanied by bond forming or bond breaking. Thus, the calculation of the Gibbs energy, see Chap. 7, is limited to reactions leaving the molecule intact, e.g. solvation processes. A completely different concept, known as "quantitative structure activity relationship" does not deal with theoretically founded energy terms. Most often, the "quantitative structure activity relationship" (QSAR) is used to predict for example structures and association constants for biochemical systems (Strasser et al. 2010a; Silva et al. 2011). Following this concept, a correlation is established between the desired property of a system and leading variables for the training systems (Kubinyi 2011). Calculating the value of the leading variables for the system of interest provides the desired property with the help of the former correlation. However, it must be emphasized that the better the system of interest corresponds to the data material representing the training set the better the prediction of binding properties will be.

Chapter 2
G Protein Coupled Receptors

The G protein coupled receptors (GPCRs) represent one of the largest families of proteins within the human genome and mediate several physiological and pathophysiological effects (Jacoby et al. 2006). GPCRs are of general interest with regard to therapy of several diseases, since about 27 % of all drugs available on market are addressing GPCRs (Fig. 2.1) (Wise et al. 2002; Overington et al. 2006).

Since a lot of literature is available with regard to GPCRs, only a short introduction is given in this chapter.

2.1 Structure of GPCRs

GPCRs are transmembrane receptors. Thus, they are located in the lipid bilayer. The GPCRs consist of seven transmembrane α helixes, spanning through the membrane from the extracellular to the intracellular part. The transmembrane domains are connected by intra- and extracellular loops. The N-terminus (amino terminus) is located on the extracellular part, whereas the C-terminus (carboxy terminus) is located on the intracellular part. Because of the structure, GPCRs are sometimes called "seven transmembrane receptors" ("7 TM receptors") (Fig. 2.2).

2.2 Different GPCR Families

GPCRs were divided into several families A–F (classes) and are described systematically (IUPHAR 2000; Fredriksson et al. 2003; Suwa et al. 2011, http://www.gpcr.org). However, there are three main families A, B and C (Table 2.1). A more detailed listing is given in the appendix GPCR Families (Source: http://www.gpcr.org/7tm).

Family A The family A GPCRs represents the largest GPCR family (IUPHAR 2000; Ballesteros et al. 2001; Chalmers et al. 2002; Jacoby et al. 2006; Mustafi et al. 2009) and is the one, which is mostly studied. The family A GPCRs, like biogenic amine receptors or (rhod)opsin (see appendix GPCR Families (Source: http://www.gpcr.org/7tm)) are the most studied so far. A disulfide bridge between the E2-loop and the upper part of TM III is typical for most of the family A GPCRs (Fig. 2.3). Additionally, most

A. Strasser, H.-J. Wittmann, *Modelling of GPCRs*,
DOI 10.1007/978-94-007-4596-4_2, © Springer Science+Business Media Dordrecht 2013

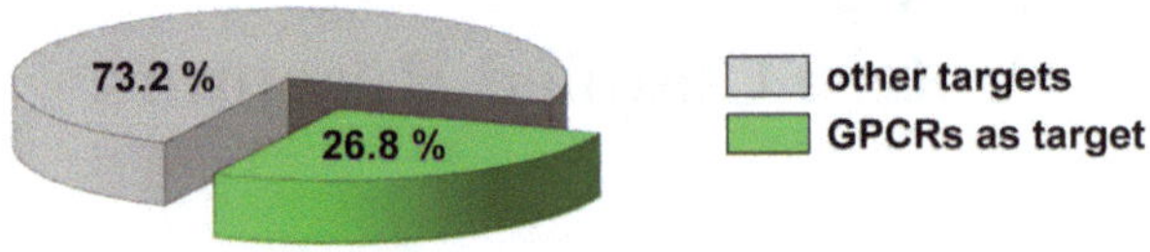

Fig. 2.1 Percentage of drugs addressing GPCRs

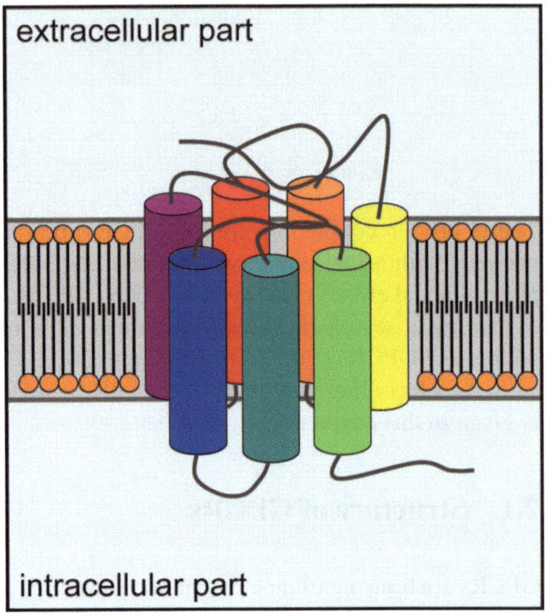

Fig. 2.2 Schematic representation if of a G protein coupled receptor, embedded in a lipid bilayer

Table 2.1 Three GPCR main families A, B and C

Family A (class I)	*Rhodopsin-like*
Family B (class II)	Secretin-like
Familiy C (class III)	Metabotropic-glutamate-like

of the family A GPCRs have a palmitoylated cysteine in the C-terminus. In general, the homology of the family A GPCRs is small. However, a small number of highly conserved key residues, like the DRY motif could be identified. Typically, small ligands of biogenic amine receptors for example, bind between the transmembrane domains of the receptor. In contrast, the binding site of peptide and glycoprotein hormone receptors is located between the N-terminus, the extracellular loops and the upper part of the transmembrane domains.

Family B GPCRs for peptides, like calcitonin, secretin or parathyroide belong to family B (IUPHAR 2000; Harmar 2001; Jacoby et al. 2006) (see appendix GPCR Families (Source: http://www.gpcr.org/7tm)). A characteristic of the family B GPCRs is the long N-terminus (Fig. 2.4). The N-terminus of family B GPCRs contains three conserved disulfide bridges (Fig. 2.4). Besides that, the extracellular loop E2 and the upper part of transmembrane domain III are connected by a disulfide bridge (Fig. 2.4). Typically, in family B GPCRs, ligands bind between the long N-terminus and the extracellular loops. Experimental data suggest that family B GPCRs prefer to couple to $G\alpha_s$ (Hoare SRJ et al. 2005).

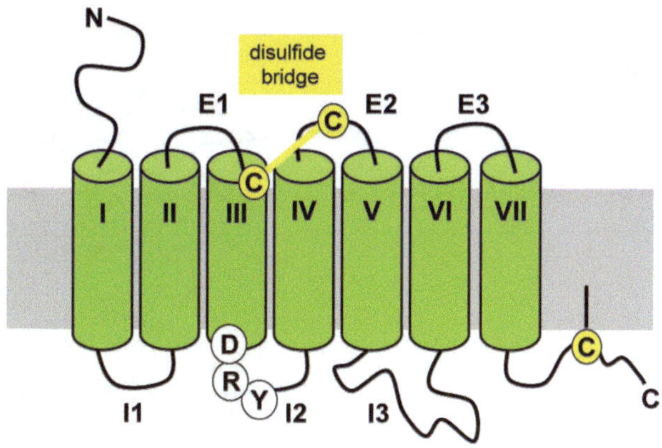

Fig. 2.3 Schematic representation of a family A GPCR

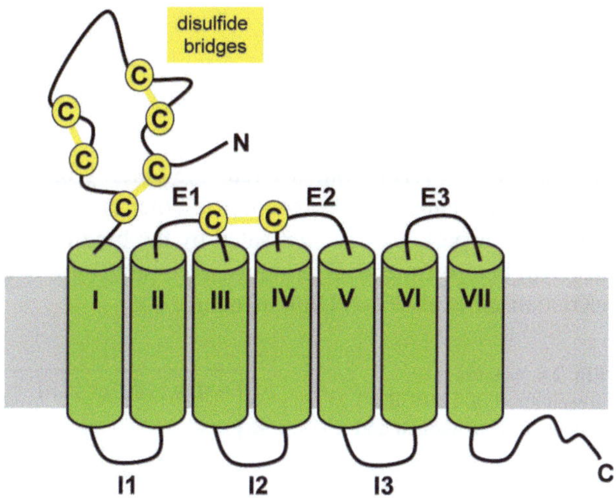

Fig. 2.4 Schematic representation of a family B GPCR

Family C Metabotropic glutamate receptors (mGluR), γ-aminobutyric acid type B ($GABA_B$) and calcium-sensing receptors (CaR) for example, belong to GPCRs of family C (IUPHAR 2000; Jacoby et al. 2006; Bräuner-Osborne et al. 2007). For most of the family C GPCRs a long N-terminus and C-terminus is typical, as well as a disulfide bridge, connecting the extracellular loop E2 with the upper part of TM III (Fig. 2.5). The ligand binding site is established by a so-called venus flytrap module (VFTM), which is connected by a cysteine-rich domain (CRD) to the transmembrane domain I.

2.3 Activation of GPCRs and Their Interaction with G Proteins

Based on several experimental data, it was shown, that GPCRs can undergo conformational changes in simplest case between an inactive and an active conformation (Kobilka and Deupi 2007). The binding of antagonists or inverse agonists stabilize

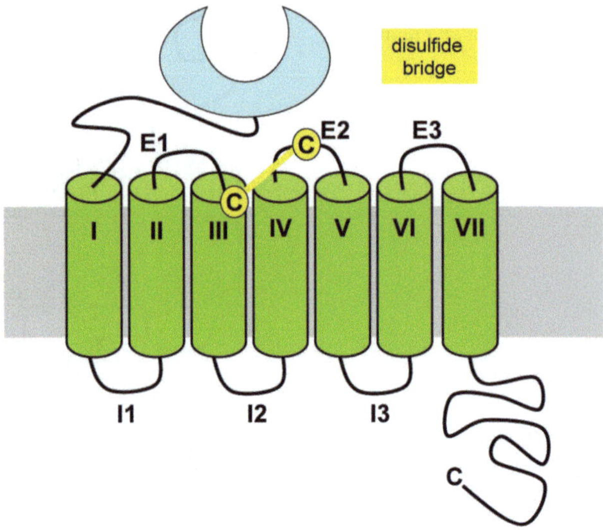

Fig. 2.5 Schematic representation of a family C GPCR

the inactive conformation, whereas the binding of (partial) agonists induce a conformational change of the receptor (Gether et al. 1998; Pierce et al. 2002). In the intracellular part, GPCRs, activated by the binding of an agonist, are able to interact with heterotrimeric G proteins, consisting of a α-, β- and γ-subunit (Fig. 2.6) (Kristiansen et al. 2004; Oldham et al. 2006).

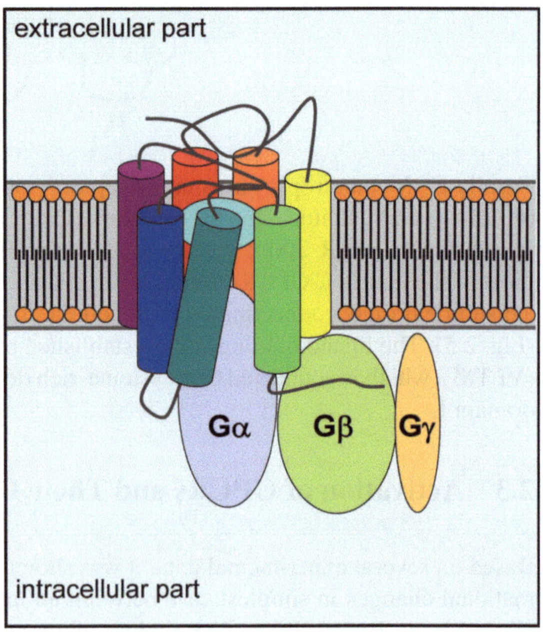

Fig. 2.6 Schematic presentation of a GPCR, activated by an agonist and interacting with a heterotrimeric G protein

There is only small knowledge about the interaction between GPCR and G protein on molecular level available. Crystal structures of GPCRs (see Chap. 3 and appendix Important Crystal Structures of GPCRs (Source: http://www.pdb.org)) on the one hand and G proteins on the other hand are known (http://www.pdb.org). In 2008, a crystal structure of opsin cocrystallized with a part of the C-terminus of Gα was published (Scheerer et al. 2008). In order to get a more detailed insight into interactions between a GPCR and the Gα-subunit, in 2010, a hβ_2R-Gα_s-complex was predicted (Strasser et al. 2010). One year later 2011, a crystal structure of the hβ_2R-G$\alpha\beta\gamma$-complex, which is shown in Fig. 2.7 was published (Rasmussen et al. 2011).

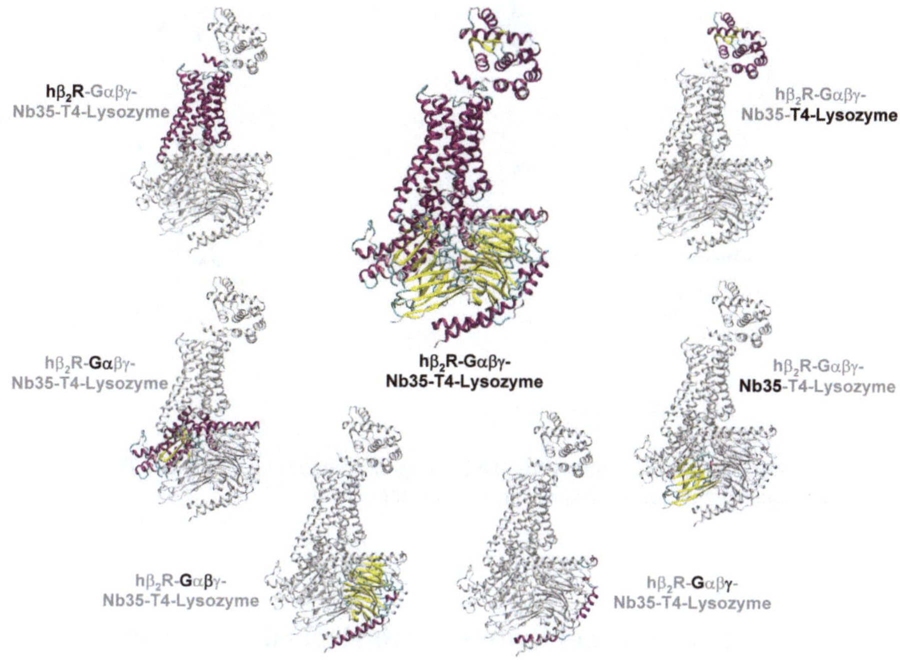

Fig. 2.7 Crystal structure of a hβ_2R-G$\alpha\beta\gamma$-Nb35-T4-Lysozyme complex. (Rasmussen et al. 2011)

G protein coupled receptors interact with heterotrimeric G proteins located in the intracellular part of a cell, comprised of a Gα-subunit and a G$\beta\gamma$ heterodimer. If an agonist binds to a GPCR, the GPCR undergoes a conformational change from the inactive to the active state (Kobilka and Deupi 2007). In the active conformation, the GPCR interacts with the appropriate G protein. Subsequently, the conformation of the Gα-subunit changes by release of GDP and a GTP binds to the ternary complex, consisting of the agonist, the GPCR and the G protein. This leads to conformational change of the Gα-subunit and the heterotrimeric G protein-complex dissociates into a Gα-GTP- and a Gβγ-complex. In dependence of the subtype of the activated Gα the appropriate signal cascades are induced selectively (Fig. 2.8) (Vauquelin and von Mentzer 2007).

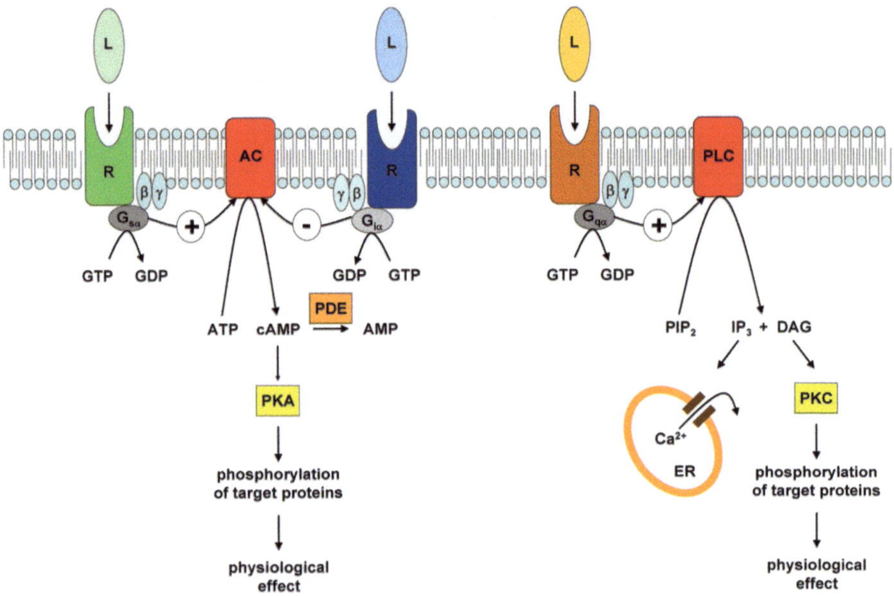

Fig. 2.8 Signalling cascade, induced by the binding of an agonist to a GPCR. Three different signalling cascades with regard to $G\alpha_s$, $G\alpha_i$ and $G\alpha_q$ are shown

2.4 Important Internet Sources with Regard to GPCRs

A very important internet source is the "GPCR network" (http://cmpd.scripps.edu) (Fig. 2.9). Here you can find important information concerning GPCRs. The "tracking status" of solving the crystal structure of distinct GPCRs might be of special interest (see Chap. 3, Fig. 3.1).

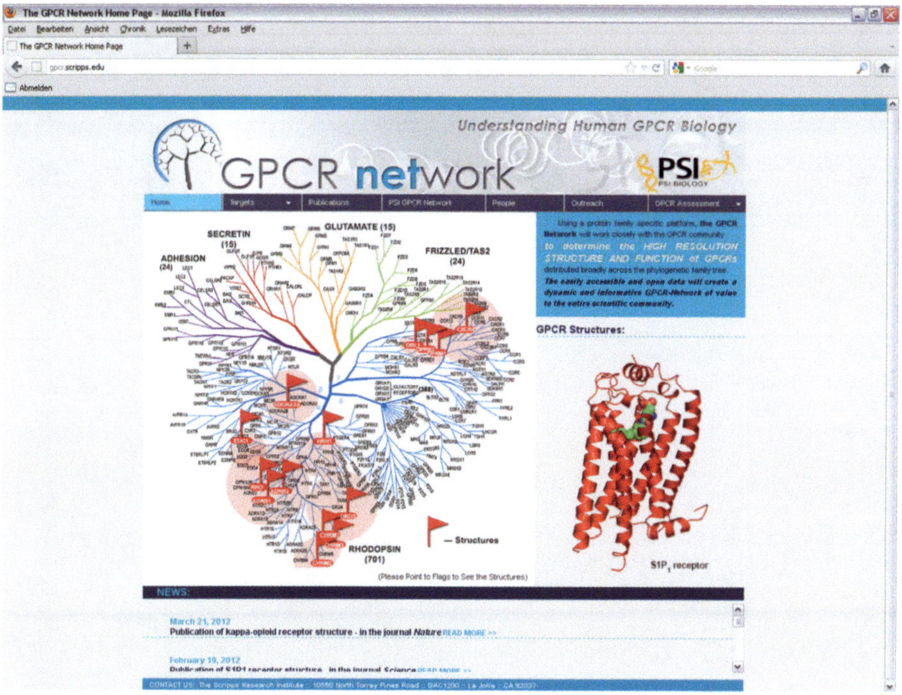

Fig. 2.9 Homepage of GPCR network. (http://cmpd.scripps.edu)

Chapter 3
Sequence Alignment and Homology Modelling

For molecular modeling of proteins in general, the structure of the protein is needed. How can such a structure be obtained? One might consider first a modeling of the protein structure *de novo* or *ab initio* based on the amino acid sequence. There are several approaches described in literature (Fleishman et al. 2006; Yarov-Yarovoy et al. 2006; Taylor et al. 2008; Zhang 2008; Barth et al. 2009; Zaki et al. 2010). For small proteins, these techniques result in suitable structures, which are in good accordance to experimentally derived structures. But it should be taken into account, that with increasing number of amino acids, thus methods are not longer appropriate, because of an exponentially increasing computational time. Thus, other techniques are necessary. One is the technique of homology modelling. This is based on the assumption that proteins of on class have a very similar structure. Thus, if the structure of one protein of a distinct class is evaluated by experimental methods, the structures of all other proteins can be modelled in homology to this experimental template. The technique of homology modelling is used with regard to several GPCRs (Zhang et al. 2006), like the NK1 receptor (Evers et al. 2004), the $P2Y_6$ receptor (Costanzi et al. 2005), the CB2 receptor (Pei et al. 2008), the NKB and NK_3 receptor (Ganjiwale et al. 2011), the cholecystokinin-1 receptor (Henin et al. 2006), histamine receptors (Jongejan et al. 2005; Preuss et al. 2007; Jongejan et al. 2008; Lim et al. 2008; Igel et al. 2009; Strasser and Wittmann 2010a; Brunskole et al. 2011) and besides addresses GPCR oligomerization (Simpson et al. 2010).

3.1 Selection of a Template

To be able to start homology modelling, one has to search for an appropriate template structure. A large number of such templates are available at the Protein Data Bank (PDB, http://www.pdb.org). Until end of 2011 a large number of crystal structures were available (Table 3.1). As illustrated by Table 3.1, most crystal structures concern the β_1- and β_2- adrenergic receptor. These crystal structures are of great interest, since different types of ligands, like inverse agonists, antagonists or (partial) agonists are bound. Thus, these crystal structures reveal important information with regard

A. Strasser, H.-J. Wittmann, *Modelling of GPCRs*,
DOI 10.1007/978-94-007-4596-4_3,

Table 3.1 pdb-codes of most important crystal structures related to opsin or GPCRs

GPCR	Related pdb-codes
Bovin (rhod)opsin	1F88, 1HZX, 1GZM, 3CAP, 3DQB, 3PQR, 3PXO
Human β_2 adrenergic receptor	2RH1, 2R4R, 2R4S, 3D4S, 3NYA, 3NY8, 3NY9, 3KJ6, 3P0G, 3PDS, 3SN6
Turkey β_1 adrenergic receptor	2VT4, 2YCW, 2YCX, 2YCY, 2YCZ, 2Y00, 2Y01, 2Y02, 2Y03, 2Y04
Human dopamine D_3 receptor	3PBL
Human histamine H_1 receptor	3RZE
Human chemokine $CXCR_4$ receptor	3ODU, 3OE6, 3OE8, 3OE9, 3OE0
Human adenosine A_{2A} receptor	3EML, 2YDO, 2YDV, 3QAK, 3PWH, 3REY, 3RFM

to different conformations of the receptors. Recently, the crystal structure of a ligand bound covalently to the hβ_2R was published (3PDS) (Rosenbaum et al. 2011). Besides the crystal structures of adrenergic receptors, 2010 the crystal structure of the human dopamine D_3 receptor (3PBL) (Chien et al. 2010) and 2011 the crystal strucuture of the human histamine H_1 receptor (3RZE) (Shimamura et al. 2011) was published. In addition to the mentioned crystal structures of biogenic amine receptors, crystal structures of the human chemokine CXCR4 receptor (Wu et al. 2010) and the human adenosine A_{2A} receptor (Jaakola et al. 2008; Lebon et al. 2011; Xu et al. 2011; Dore et al. 2011) are known (Table 3.1).

Thus, if a GPCR has to be modelled an appropriate template has to be chosen. If one likes to model a biogenic amine receptor by homology modelling, the crystal structure of a biogenic amine receptor is suggested to be used as template to solve this task. For modelling of inverse agonists or neutral antagonist in the receptor bound state, a template, representing the inactive conformation should be chosen, whereas a template, representing the active conformation should be used in case of (partial) agonists. Furthermore, the homology between the receptor to be modelled and the template should be as high, as possible. Based on these suggestions, it is the responsibility of the modeller to choose an appropriate template for homology modelling.

Sometimes, a look onto the homepage of GPCR network (http://cmpd.scripps.edu) is very useful. There, you get information about the tracking status of GPCRs which will be crystallized in future (Fig. 3.1).

3.2 Crystal Structures of GPCRs (Source: http://www.pdb.org)

In the appendix, the most important information with regard to all crystal structures of (rhod)opsin or GPCRs is summarized tabular. These tables should give you a fast overview onto available crystal structures, resolution, structure of a cocrystallized ligand, related UniProtKB entries and corresponding literature. Have a careful look onto the section "mutation"! Often, not the wild type receptor is crystallized, instead point mutations were introduced. Thus, if you want to model the receptor, which is crystallized, you may change the amino acids, mutated in the crystal structure, into the corresponding amino acid of the wild type receptor. An overview of the differences in crystal structures is given by the Figs. 3.2–3.6.

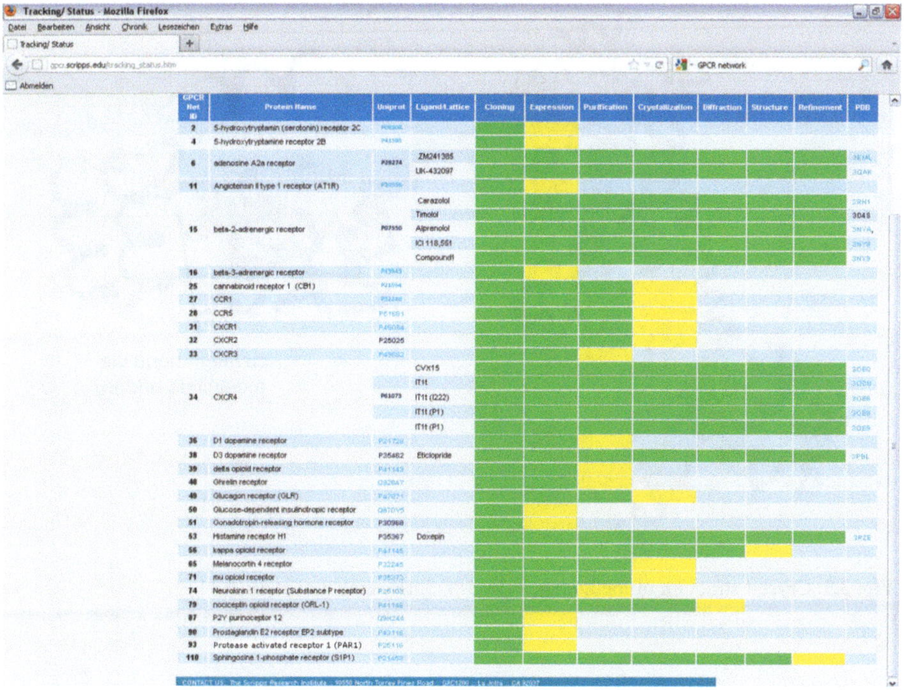

Fig. 3.1 GPCR tracking status. (Status: November 2011; Source: http://gpcr.scripps.edu/tracking_status.htm)

Fig. 3.2 Crystal structure of the turkey β_2R, 2Y00. (Warne et al. 2011)

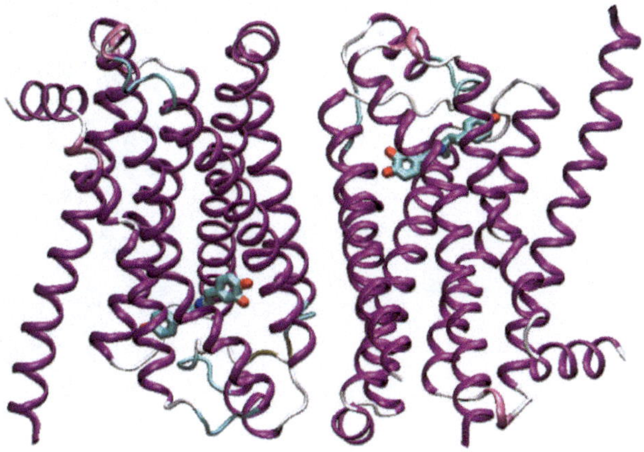

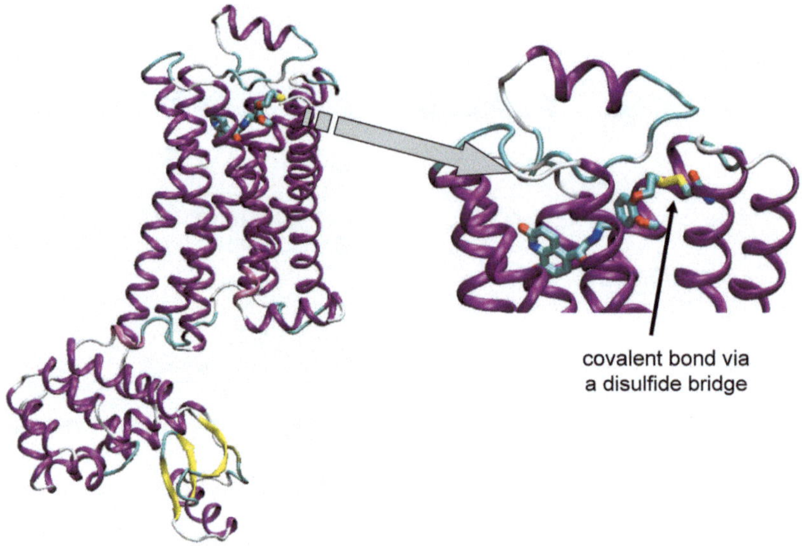

Fig. 3.3 Crystal structure of the human β_2R, 3PDS. (Rosenbaum et al. 2011)

Fig. 3.4 Crystal structure of the human CXCR4, 3ODU. (Wu et al. 2010)

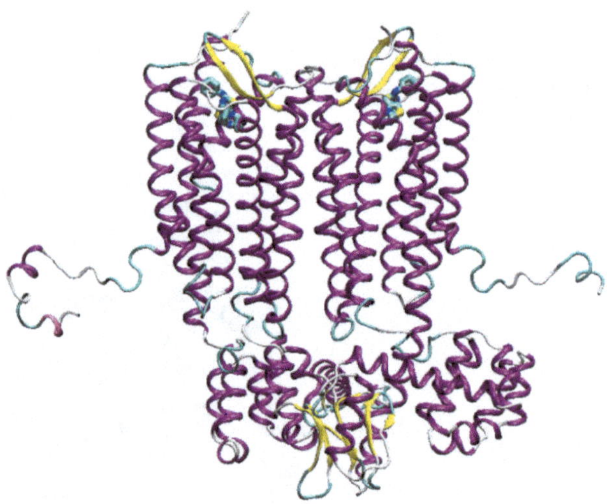

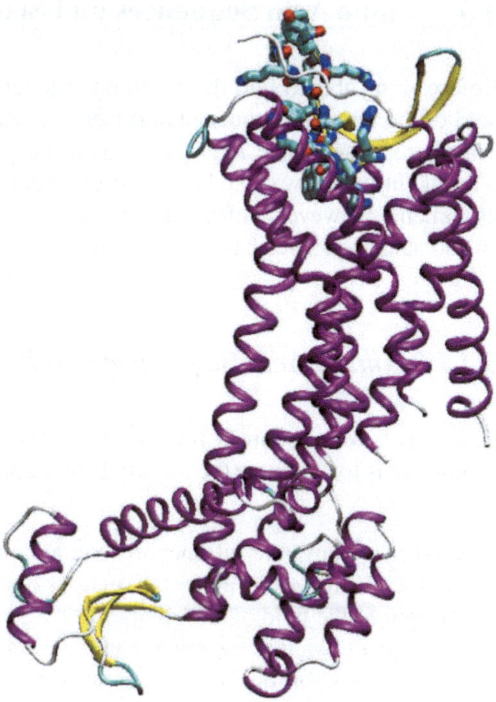

Fig. 3.5 Crystal structure of the human CXCR4, 3OE0. (Wu et al. 2010)

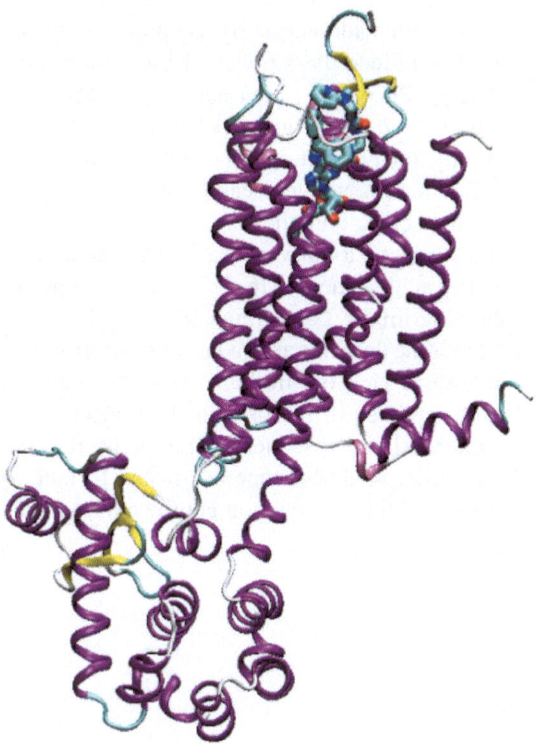

Fig. 3.6 Crystal structure of the human $A_{2A}R$, 3EML. (Jaakola et al. 2008)

3.3 Amino Acid Sequences and Sequence Alignment

Before being able to start the homology modelling, it has to be decided which amino acid of the template sequence corresponds to an amino acid in the target sequence. Therefore, a sequence alignment has to be performed manually or automatically. Clustal (http://www.clustal.org) for example, is a software for multiple sequence alignment. However, before starting with sequence alignment, the corresponding amino acid sequences have to be obtained.

3.3.1 Amino Acid Sequences – Where to Get From?

There are several sources for amino acid sequences present in the internet. One prominent is for example the Expasy Proteomics Server (http://expasy.org) (Fig. 3.7).

Exercise Start your internet browser and open the site http://expasy.org. Now choose "UniProtKB" under the section "query". Then you can type your search string into the field on the right.

Now we want to search for the human adrenergic β_2 receptor. There are different possibilities for the search string. For example, type "adrenergic" and click the "Search" button. Now, more than 900 results, related to "adrenergic" are presented. Scroll, until the receptor of your choice is listed. In our case it is the human adrenergic β_2 receptor with the accession code "P07550". If you want to reduce the number of hits, the search string has to be defined more exactly. Please try "beta adrenergic receptor", "beta-2 adrenergic receptor" and "beta-2 adrenergic receptor human". By defining the search string more exactly, the number of hits can be significantly reduced and it is easier for you to find the hit, you are searching for.

Now, click, onto the corresponding entry with the accession code "P07550" and you get a lot of very useful information about this receptor, including the amino acid sequence. In the section "Regions", the amino acids, related with the N-terminus, C-terminus, intracellular loops, extracellular loops and trans-membrane domains are given. This information is very helpful for the sequence alignment later on. In the section "Sequence" you can find the whole amino acid sequence of the protein. For further proceeding on with the amio acid sequence like for sequence alignment, it may be easier for you, to download the amino acid sequence as "fasta" format. To do so, please click onto the string "FASTA". Now you get the amino acid sequence as simple ascii file.

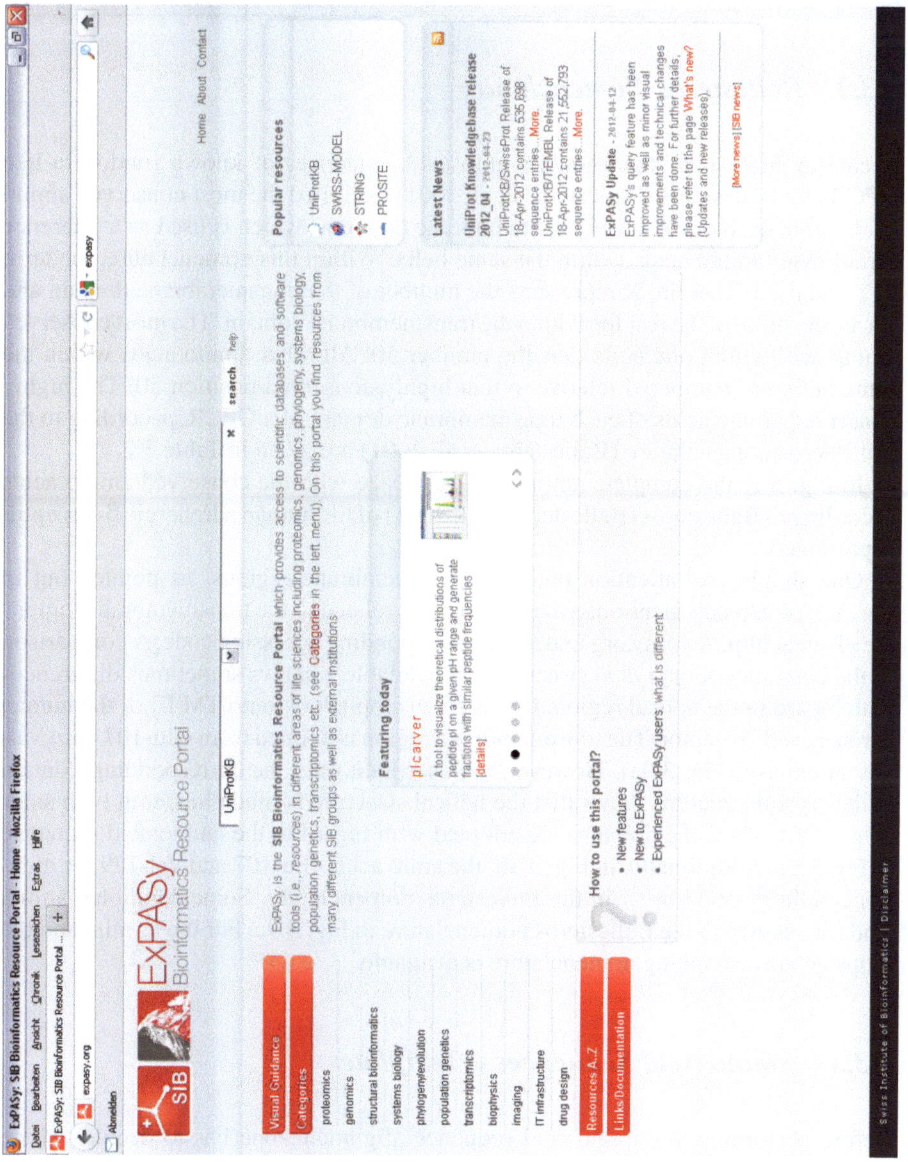

Fig. 3.7 Homepage of the expasy server. (http://expasy.org)

Table 3.2 Highly conserved amino acid according to Ballesteros (Ballesteros et al. 2001) of each transmembrane domain of rhodopsin-like GPCRs

TM I	TM II	TM III	TM IV	TM V	TM VI	TM VII
Asn, N	Asp, D	Arg, R	Trp, W	Pro, P	Pro, P	Pro, P

3.3.2 Ballesteros Nomenclature

A careful analysis of the known amino acid sequences of known rhodopsin-like GPCRs by Ballesteros (Ballesteros et al. 2001) exhibited the most conserved amino acid within each of the seven transmembrane domains, which is used as a reference for all other amino acids within the same helix. Within this nomenclature, the term X.YY is used. Therein, X represents the number of the transmembrane domain and YY the position of the residue within the transmembrane domain. The most conserved amino acid within one helix gets the number 50. All other amino acids within the same helix are numbered relative to that highly conserved position 50. The highly conserved amino acids of each transmembrane domain of a GPCR, according to the Ballesteros nomenclature (Ballesteros et al. 2001) are given in Table 3.2.

In Fig. 3.8, the complete amino acid sequence with the conserved amino acids according to Ballesteros (Ballesteros et al. 2001) of the human adrenergic β_2 receptor is presented.

One should pay attention onto the transmembrane regions, as pointed out in Fig. 3.8. As already mentioned the amino acids related to the transmembrane regions are given at http://expasy.org under the corresponding accession code. A comparison to the corresponding crystal structure – if available – shows sometimes differences with regard to the helical region. Let us for example look onto TM III of the human adrenergic β_2 receptor. The transmembrane region is defined from Glu-107 until Val-129 at expasy (Fig. 3.9a). However, a closer look onto the corresponding domain at the crystal structure shows that the helical structure is much longer at both sides (Fig. 3.9b). Thus, the domains are adopted with regard to the amino acid sequence in Fig. 3.9c. Additionally, in Fig. 3.9b, the amio acids Glu-107 and Val-129 are mentioned $Glu^{3.26}$ and $Val^{3.48}$ in the Ballesteros nomenclature. Some additional amino acids are shown in the Ballesteros nomenclature in Fig. 3.9c. For the termini and the loops no corresponding nomenclature is available.

3.3.3 Amino Acid Sequences – Templates

Before performing an amino acid sequence alignment, one has to decide, which structure should be used as template structure for homology modelling. Meanwhile a lot of crystal structures of bovin rhodopsin or GPCRs like the human adrenergic β_2 receptor or turkey adrenergic β_1 receptor are available (see Tab. 3.1 and appendix Important Crystal Structures of GPCRs (Source: http://www.pdb.org)). It cannot be decided overall, which crystal structure should be used as a template for

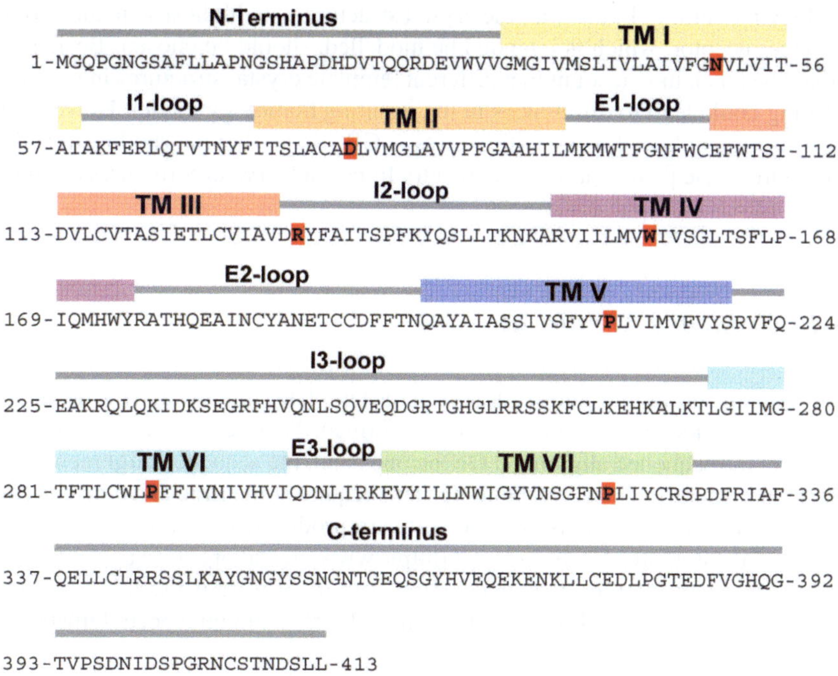

Fig. 3.8 Amino acid sequence of the human adrenergic β_2 receptor. The transmembrane domain are presented, as defined at http://expasy.org, accession code P07550. The highly conserved amino acids, defined by Ballesteros (Ballesteros et al. 2001) are marked by *red boxes*

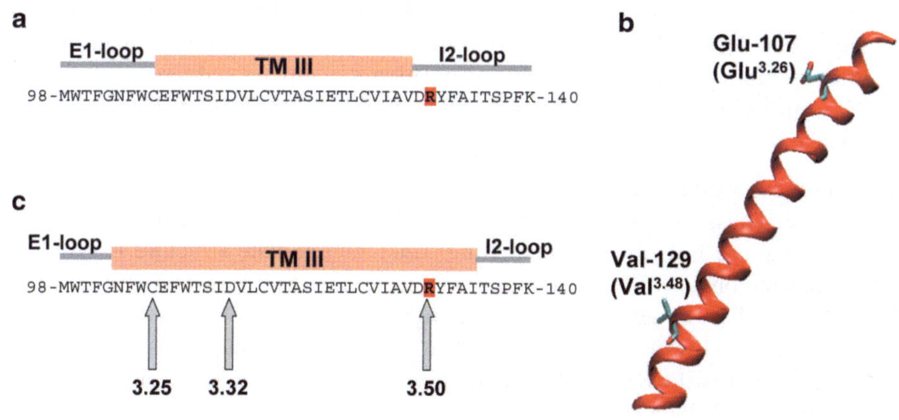

Fig. 3.9 Helical structure of a transmembrane domain. **a** Definition of the TM domain III of the human adrenergic β_2 receptor at expasy (http://www.expasy.org). **b** TM III of the human adrenergic β_2 receptor of a crystal structure. **c** Amino acid sequence of TM domain III, based on the crystal structure

homology modelling. In general, the crystal structure with highest sequence homology to the receptor, which is intended be modelled, should be chosen. Besides that it should be taken into account that different template crystal structures in homology modelling could lead to differences in the resulting homology model. However, the mainly used templates for modelling class A GPCRs are bovine rhodopsin and the human adrenergic β_2 receptor (see appendix Important Crystal Structures of GPCRs (Source: http://www.pdb.org)).

3.3.4 Sequence Alignment

After retrieving the amino acid sequences of the template structure and the destination receptor, the sequence alignment can be performed. There exist several techniques, to perform the sequence alignment. On the one hand, the sequence alignment can be performed manually. The corresponding steps require some time and concentration. On the other hand, there exist several software products, which allow performing an alignment automatically, like clustal (http://www.clustal.org) (see appendix Summary of Important Internet Resources). However, if software is used, it is definitely necessary to check to resulting alignment in order to avoid unexpected mistakes or some inaccuracies.

For a manual sequence alignment, the alignment is performed by several steps:

1. Use the information of the expasy server (http://expasy.org) to get an idea about the amino acids of the seven transmembrane domains for template and target sequence.
2. Perform the sequence alignment for each transmembrane domain in ascending order. Here, it is necessary, that the highly conserved amino acid of each transmembrane domain has the same position in template and target.
3. Now, the alignment for the termini and loops can be performed. There you have to take into account several points:

   - In most crystal structures, the N-terminus and C-terminus are often not complete. Thus, there you can perform the alignment of such regions, but there is no real use in homology modelling, since no template structure is given for such regions.
   - The I1-, E1-, I2- and E3-loop can be aligned easily in most cases to the template sequence. However it should be taken into account, that corresponding loops of different GPCRs could differ in their length. This has to be taken carefully into account later on in the homology modelling. To declare a vacant position in amino acid sequence, a hyphen (-) is used in general.
   - The I3-loop differs significantly in length (from some ten to some hundred amino acids) within the different GPCRs. Additionally, the I3-loop is not completely present in the crystal structures, available up to now. Thus, a complete I3-loop alignment is useless for homology modelling. However, for MD simulations, it will be useful to close the open ends between TM V and TM VI.

```
             N-Terminus                        TMI               I1          TMII
hH1R   1--------MSLPNSSCLLEDKMCEGNKTTMASPQLMPLVVVLSTICLVTVGLNLLVLYAVRSERKLHTVGNLYIVSLSVADLIVGA-78
hH4R   1--------------------MPDTNSTINLSLSTRVTLAFFMSLVAFAIMLGNALVILAFVVDKNLRHRSSYFFLNLAISDFFVGV-66

                 E1                TMIII                  I2                   TMIV
hH1R  79-VVMPMNILYLLMSKWSLGRPLCLFWLSMDYVASTASIFSVFILCIDRYRSVQQPLRYLKYRTK-TRASATILGAWFLSFLWVIPI-162
hH4R  67-ISIPLYIPHTLFE-WDFGKEICVFWLTTDYLLCTASVYNIVLISYDRYLSVSNAVSYRTQHTGVLKIVTLMVAVWVLAFLVNGPM-150

               E2                   TMV                         I3
hH1R 163-LGW---NHFMQQTSVRREDKCETDFYDVTWFKVMTAIINFYLPTLLMLWFYAKIYKAVRQHCQ....QYVSGLHMNRERKAAKQL-417
hH4R 151-ILV----SESWKDEG--SE-CEPGFFSEWYILAITSFLEFVIPVILVAYFNMNIYWSLWKRDH....HQREHVELLRARRLAKSL-305

            TMVI              E3             TMVII                 C-Terminus
hH1R 418-GFIMAAFILCWIPYFIFFMVIAFCKNC--CNEHLHMFTIWLGYINSTLNPLIYPLCNENFKKTFKRILHIRS-------------487
hH4R 306-AILLGVFAVCWAPYSLFTIVLSFYSSATGPKSVWYRIAFWLQWFNSFVNPLLYPLCHKRFQKAFLKIFCIKKQPLPSQHSRSVSS-390
```

Fig. 3.10 Manual alignment of the hH_4R to the hH_1R. *green*: termini and loops; *grey*: transmembrane domains; *red boxes*: highly conserved amino acids (Ballesteros et al. 2001); *yellow*: highly conserved cysteine, establishing a disulfide bridge to the upper part of TM III; -: missing amino acids; the amino acids of the I3-loop are not shown completely, which is indicated by *dots*

Therefore, some amino acids of the beginning and end of the I3-loop are modelled correctly and the gap is closed by an alanine chain.

- The E2-loop has to be aligned very carefully. It has to be taken into account, that there is a highly conserved disulfide bridge between the E2-loop and the upper part of TM III. Thus, the corresponding cysteine has to be positioned correctly.

An example for an alignment of the human histamine H_4 receptor to the human histamine H_1 receptor is shown in Fig. 3.10.

3.4 Homology Modelling

3.4.1 Modelling of the Transmembrane Domains

The helical transmembrane domains can be easily modelled straight forward. Therefore, only the amino acid side chains have to be changed into the side chain of the destination with appropriate modelling software.

3.4.2 Modelling of Loops

In general the transmembrane domains of different GPCRs consist of the same number of amino acids. Thus, the homology modelling of transmembrane domains is quite easy and can be performed straight forward. In case of intra- or extracellular loops, which are connecting the transmembrane domains, differences in number of amino acids of a loop between different GPCRs can occur. This is the case for the E2- or E3-loop between hH_1R and hH_4R (Fig. 3.10). Small gaps can be closed with "loop search" modules by using appropriate software. For some biogenic amine

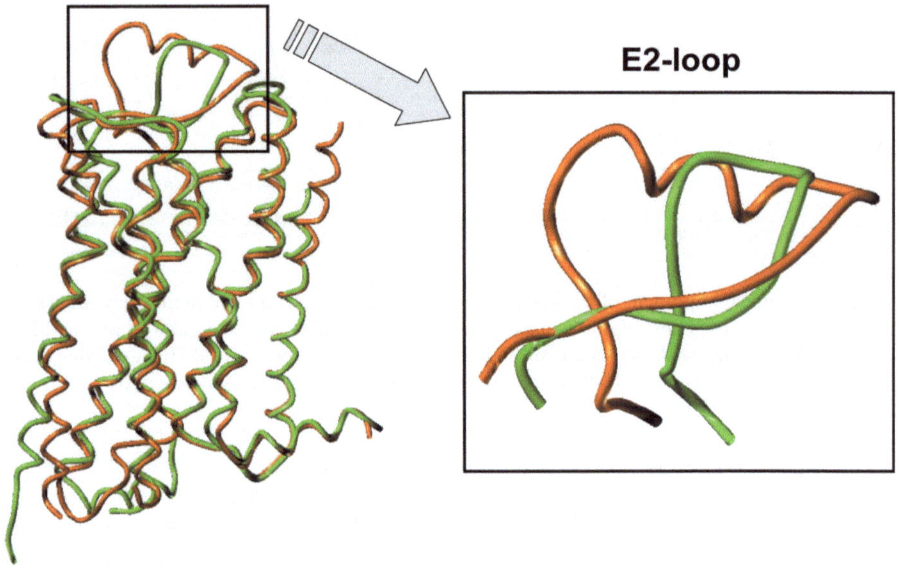

Fig. 3.11 Different conformations of the E2-loop based on crystal structures

receptors, an influence of extracellular loops, especially the E2-loop, onto the binding of ligands to the receptor was shown (Lim et al. 2008; Brunskole et al. 2011). Thus, a correct modelling of the loops is very important. Most of the loops are resolved by crystal structures. However, this is often not the case with regard to the extracellular loop E2 and this is not the case with regard to the intracellular loop I3.

Since the E2-loop is in contact with the binding pocket, the E2-loop has to be modelled completely. If you look onto different crystal structures with complete E2-loop, you can see different conformations (Fig. 3.11).

Thus, you have to decide carefully, which template is to be used for modelling of the E2-loop. A large number of crystal structures are obtainable for the human adrenergic hβ_2 receptor. But the hβ_2R is a special case: There are two disulfide bridges in the E2-loop (Fig. 3.12), whereas in most others GPCRs there is only one disulfide bridge in the E2-loop, connecting the E2-loop with the upper part of the TM III.

A part of the E2-loop of the hβ_2R exhibits a helical structure, but this is not the case for all other GPCRs. Thus, you have to decide carefully, if it would be appropriate to use two different template structures for homology modelling: one for the E2-loop and one for the remaining parts of the receptor. However, the E2-loops are widely different in their length, thus, in most cases, the E2-loop cannot be modelled by changing an amino acid side chain of the template into the side chain of the destination. Thus, you have to use also techniques, like "loop search". For only one loop search, the number of amino acids is too long, and you would get bad results. Thus, it is better, to use at least one fixed point. This is the highly conserved cysteine, connecting the E2-loop by a disulfide bridge with the upper part of TM III (Fig. 3.10).

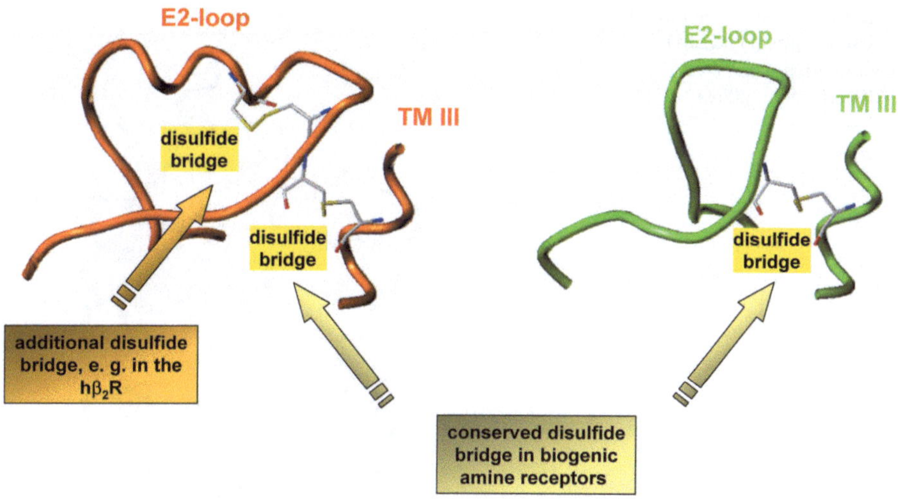

Fig. 3.12 Different numbers of disulfide bridges in the E2-loop

3.4.3 Modelling of Internal Water

A detailed analysis of the crystal structures of GPCRs reveals that there are internal, highly conserved water molecules present (Fig. 3.13). Several studies showed that these water molecules are involved in the hydrogen bonding within the receptor. Based on the published data, it can be suggested, that these water molecules are essential for stabilizing the receptor or important for receptor activation (Pardo et al. 2007). Thus, in order to generate a stable receptor model, the water molecules which are localized/crystallized within the receptor should be included into the homology model.

3.4.4 Modelling of the C-Terminal Part of the Gα Subunit or the Whole Gα Subunit

Based on several studies it is suggested, that a GPCR in its active conformation interacts in the intracellular part with the Gα subunit. There is only small knowledge about the receptor – G protein interaction. However, recently, the crystal structure of opsin, cocrystallized with eleven amino acids of the C-terminus of the Gα subunit (Scheerer et al. 2008) and a complete GPCR – G protein complex (Rasmussen et al. 2011) were published. A detailed analysis of the corresponding crystal structures (3DQB, 3SN6) shows, that the C-terminus of the Gα subunit is deeply bound in a pocket between the transmembrane domains. Leaving out this part of the Gα will result in some problems in subsequent molecular dynamic simulations. In general, if molecular dynamic simulations of a receptor are performed, the receptor is embedded in its natural surrounding. Thus, if the C-terminal part of Gα or the whole Gα is

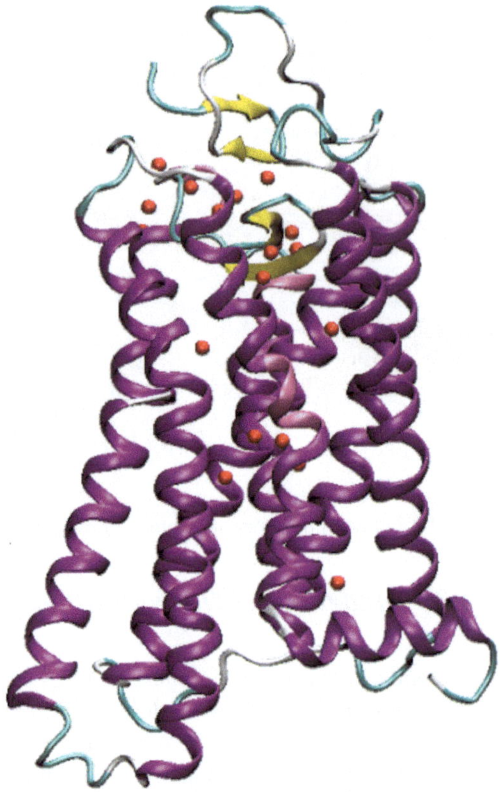

Fig. 3.13 Crystal structure of bovin rhodopsin (1GZM) with internal water (*red balls*). (Li et al. 2004)

missing, the resulting free space is filled with water molecules. Water molecules are highly polar and thus have completely other (surface) properties than the C-terminal part of Gα. Thus, leaving out the C-terminal part of Gα and substitution by water molecules in molecular dynamics can lead to instabilities of the receptor during the molecular dynamic simulation. Thus, it is suggested, to include at the whole Gα or least the C-terminal part of Gα in a homology model. Be aware, that each GPCR couples to a distinct Gα subunit (Fig. 2.8).

3.4.5 Refinement of the Receptor Model

After finishing the homology modelling, several checks of the complete model should be performed. A typical error of beginners in molecular modelling is presented in Figs. 3.14–3.16. During homology modelling, some amino acid side chains have to be mutated into the correct amino acid side chain. Sometimes, especially with regard to long side chains or aromatic rings, collisions between the side chains arise. There are two types of collisions: In the first type, two side chains are in close contact, as shown in Fig. 3.14. In most of these cases, energy minimization is sufficient to

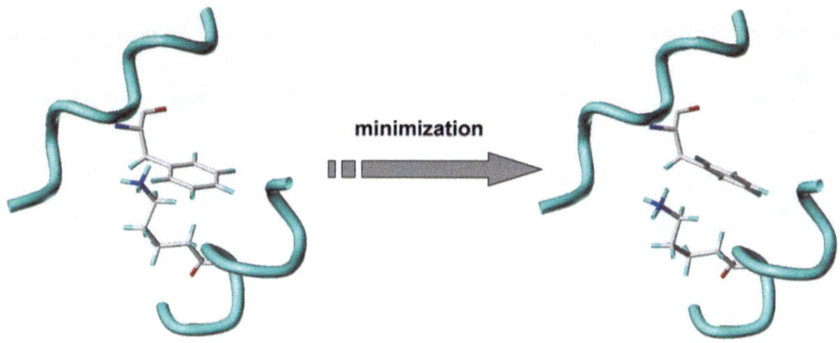

Fig. 3.14 Close contact between the atoms of a lysine and phenylalanine: *Left*: before minimization, *right*: after minimization

Fig. 3.15 Part of a protein structure after minimization. What is the problem?

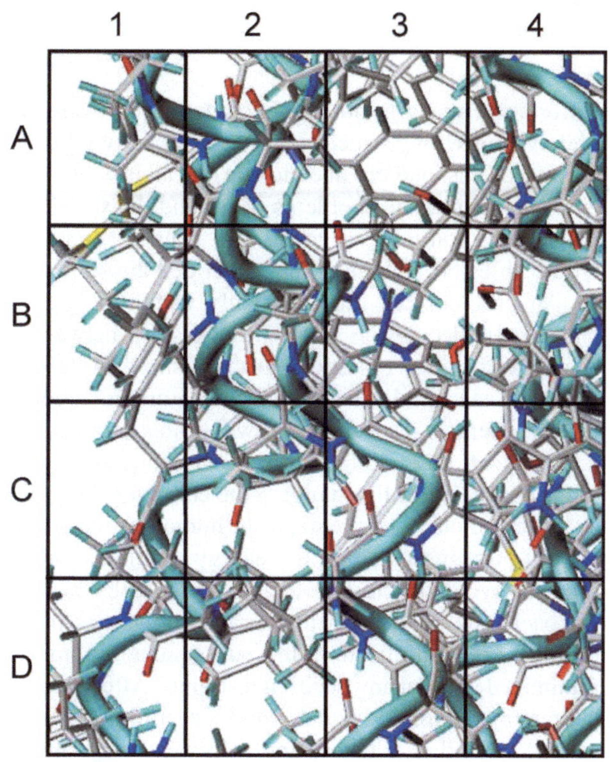

remove the collision and suitable structures might be obtained. The second type of collision is a more difficult pitfall, which is illustrated in Figs. 3.15 and 3.16. Look carefully onto the Fig. 3.15. Where is the problem?

After a careful look onto the picture you may see, that there is a problem with regard to a lysine and phenylalanine in box B3. This is also illustrated in Fig. 3.16.

Here, a long amino acid side chain, like present in lysine, is located within an aromatic ring, like present in tyrosine, phenylalanine, tryptophane or histidine.

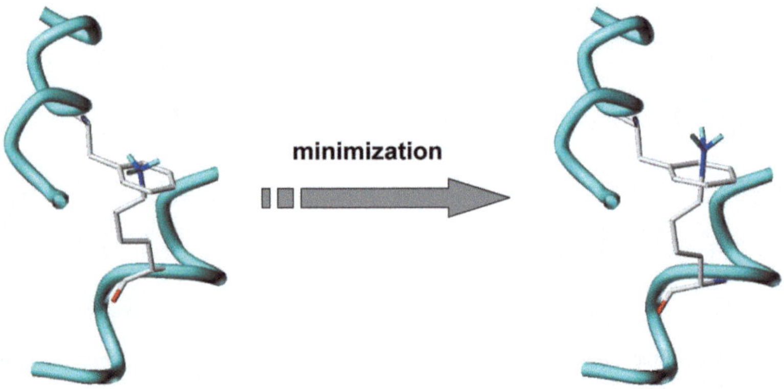

Fig. 3.16 Wrong close contact between the atoms of a lysine and phenylalanine: *Left*: before minimization, *right*: wrong structure after minimization

Unfortunately, a large number of modelling software minimizes a protein, containing such type of wrong structure. And additionally in most cases the potential energy is negative. Thus, one might conclude that all is well. However, often during molecular dynamic simulation, problems occur and the simulation stops with an error. If this is the case, you have to go back to your starting structure and look for the error. Often, an error, similar to that described above (Fig. 3.16) causes the problem. A similar problem can occur not only within the protein, but also between protein and lipid molecules. If there are collisions between amino acid side chains, one has to decide, how to remove this collisions. In general, there are two possibilities: First, one can simply perform an energy minimization. But in some cases, this could lead to artefacts, especially, if two aromatic moieties are linked together. Thus, it is suggested, that one looks separately onto each collision and tries to remove the collisions by carefully changing the corresponding dihedral angles.

After completing these steps, the homology model can be energetically minimized. Here it is suggested, that the energy minimization is performed step by step. In order to avoid structural artefacts, induced by minimization, it is important, that the backbone of the transmembrane domains is provided with position restraints during a first minimization. In a subsequent minimization steps, the receptor can be minimized without any position restraints. Afterwards, the model should be checked, addressing the following items and if everything is correct, one can start with further modelling studies, like docking or molecular dynamic simulations.

- ☐ Check for the correct amino acid sequence
- ☐ Check for the presence of the disulfide bridge between the E2-loop and the upper part of TM III
- ☐ Check for the correct configuration of the amino acids
- ☐ Check for collisions or bad contacts between amino acid side chains

Chapter 4
Construction of Ligands

Some molecular modelling software include very comfortable editors, which allow to construct ligands. Additionally, distinct atom types can be assigned to these atoms. In contrast, Gromacs (http://www.gromacs.org) is a powerful software package for molecular dynamic simulations and no editor for construction of ligands is included. Therefore, we recommend that you download an appropriate editor, like chimera (http://www.cgl.ucsf.edu/chimera/) and install this on your computer. In general, you can also use other software to construct your 3D-molecules, but the software should be able to save the molecule as pdb-file. Before you can use Gromacs (http://www.gromacs.org) to simulate organic compounds, like a ligand, you have to generate a topology-file of the molecule of interest. In general, your molecule editor is not able to create an appropriate topology-file. We do not recommend constructing a topology-file manually, because therefore you need detailed knowledge about types of the atoms or sites and their force field parameters on the one hand. On the other hand, you have to define bonds, angles and dihedrals, which is a very complicated procedure for a beginner. Besides, generating a topology file manually, is very susceptible for mistakes. Solving this task, you can use the PRODRG-server (http://davapc1.bioch.dundee.ac.uk/prodrg/). The starting page of the server is shown in Fig. 4.1.

If you have started the PRODRG server (http://davapc1.bioch.dundee.ac.uk/prodrg/), use the button "Run PRODRG", which will bring up the next page (Fig. 4.2):

The academic use of the PRODRG server is free, but in order to avoid abuse, one user is allowed to perform about three runs per day. Therefore, you have to order a so-called "token" by submitting your e-mail address. Within some minutes, you should get your "token". Now copy and paste your valid "token" into the appropriate field. Afterwards, you can fill the remaining fields and submit your PRODRG job. For obtaining the GROMACS coordinate and topology-file of ethanol for example, start a molecule editor, like chimera, construct ethanol and save the molecule as pdb-file, named `ethanol.pdb`.

A. Strasser, H.-J. Wittmann, *Modelling of GPCRs*,
DOI 10.1007/978-94-007-4596-4_4,

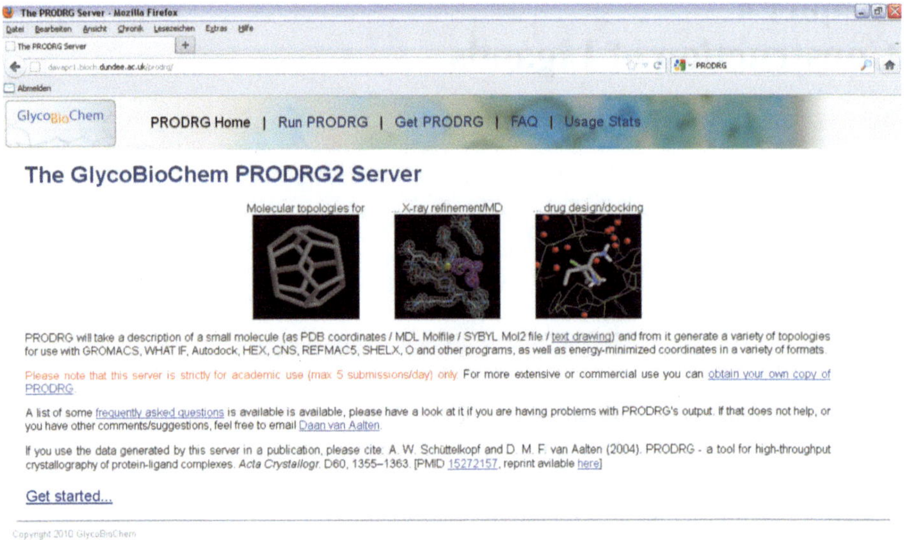

Fig. 4.1 Homepage of the PRODRG server. (http://davapc1.bioch.dundee.ac.uk/prodrg/)

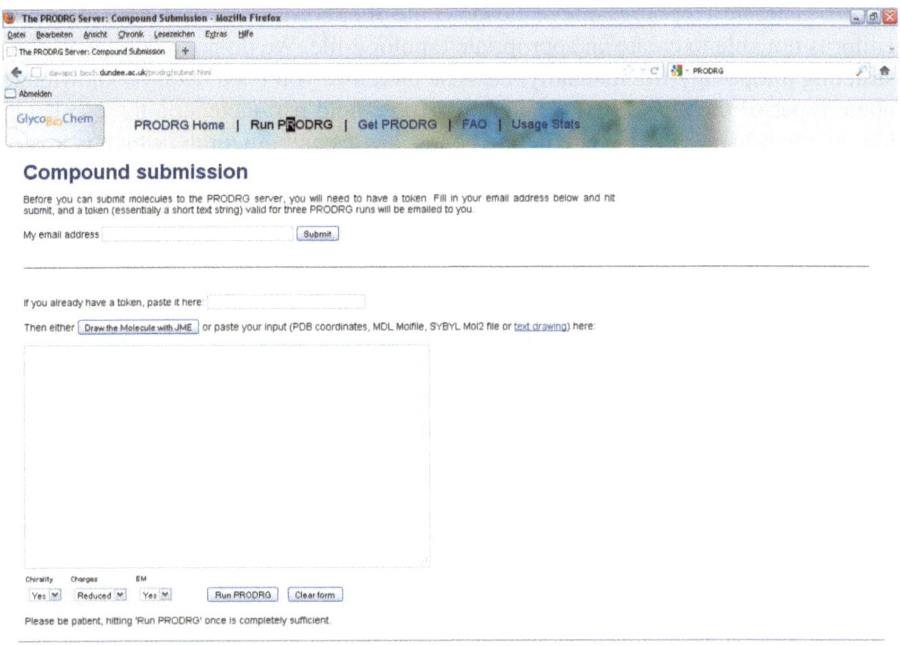

Fig. 4.2 Site for compound submission of the PRODRG-server

```
HEADER ETHANOL
COMPND ETHANOL
REMARK GENERATED BY SYBYL (TRIPOS, INC.) 15-AUG-10
HETATM 1  O2 LIG  1  -8.207   3.565 -0.317 1.00   -0.40
HETATM 2  H3 LIG  1  -7.452   3.455  0.253 1.00    0.21
HETATM 3  C1 LIG  1  -9.325   2.747  0.067 1.00    0.04
HETATM 4  H2 LIG  1  -9.665   3.010  1.084 1.00    0.06
HETATM 5  H1 LIG  1  -10.148  2.959 -0.635 1.00    0.06
HETATM 6  C4 LIG  1  -8.958   1.244 -0.007 1.00   -0.04
HETATM 7  H6 LIG  1  -8.620   0.992 -1.024 1.00    0.03
HETATM 8  H5 LIG  1  -8.149   1.012  0.702 1.00    0.03
HETATM 9  H4 LIG  1  -9.834   0.624  0.241 1.00    0.03
CONECT 1    2    3
CONECT 2    1
CONECT 3    1    4  5  6
CONECT 4    3
CONECT 5    3
CONECT 6    3    7  8  9
CONECT 7    6
CONECT 8    6
CONECT 9    6
MASTER 0    0    0    0    0    0    0    0    9    0    9    0
END
```

Open the file `ethanol.pdb` with an editor, copy all data and paste them into the appropriate field of the PRODRG-site, shown in Fig. 4.2. Since ethanol is not chiral, you can choose "no" at the corresponding field. Additionally choose "full charges" and "yes" with regard to EM (EM means energy minimization). Afterwards, click at the button "Run PRODRG". After some minutes, you obtain the results-page. At first, you see some remarks of the server and additionally, the molecule with (added) hydrogens is shown (Fig. 4.3).

If your scroll down, you see a summary of different output-files (Fig. 4.4). Most important, concerning GROMACS is the third item under "Coordinates" and the first item under "Docking/MD simulations".

Within the "Coordinates" section for GROMACS, you find three different items, namely a coordinate file with "polar hydrogens", with "polar/aromatic hydrogens" and with "all hydrogens". If you look onto the number of coordinate lines you see differences, in case of ethanol, between "polar hydrogens" and "all hydrogens". Since the site-concept is used in GROMACS, the hydrogens of an alkyl-moiety are integrated within the carbon. This means for example, that a methyl group (CH_3) does not consist of four sites – one carbon and three hydrogens – instead, it is summarized in one site (see Chap. 9). This is a very important aspect with regard to simulation time. Because of the combination of several atoms to one site, the number of sites is reduced and this leads to an exponential decrease in simulation time. If you compare with the contents of the topology-file, the coordinate file "polar/aromatic" hydrogens is relevant. Be aware, that the number of coordinates in the gro-file (Fig. 4.5) has

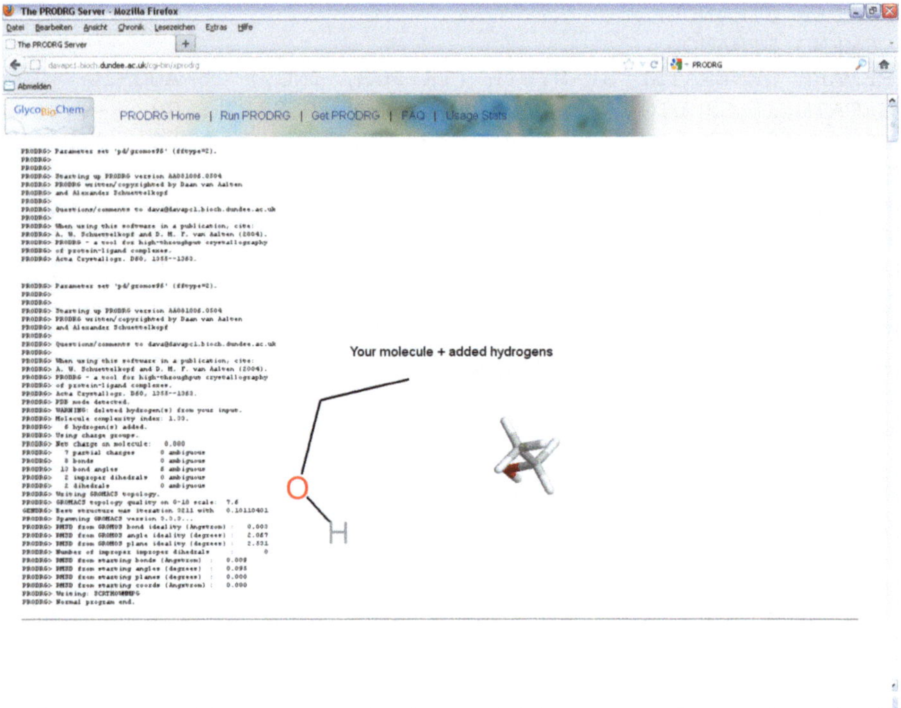

Fig. 4.3 First part of the results-page of the PRODRG-Server

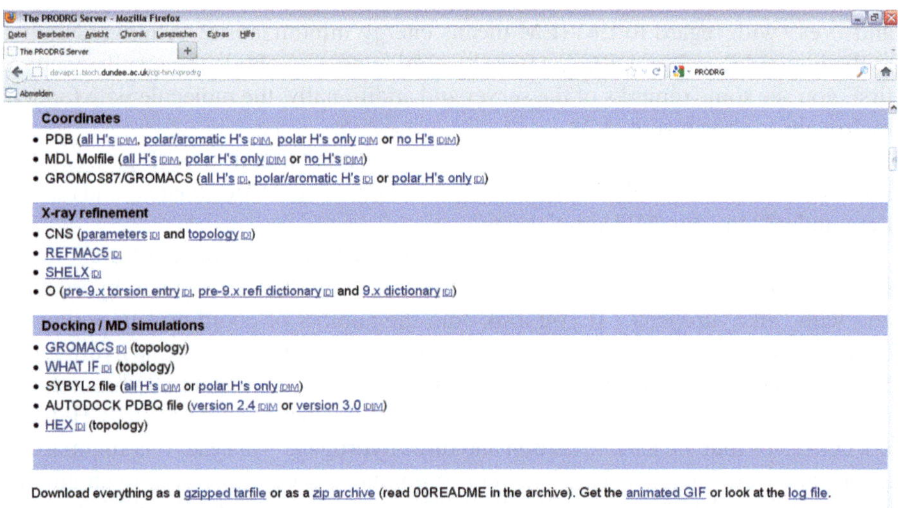

Fig. 4.4 Overview of different outputs of the PRODRG server

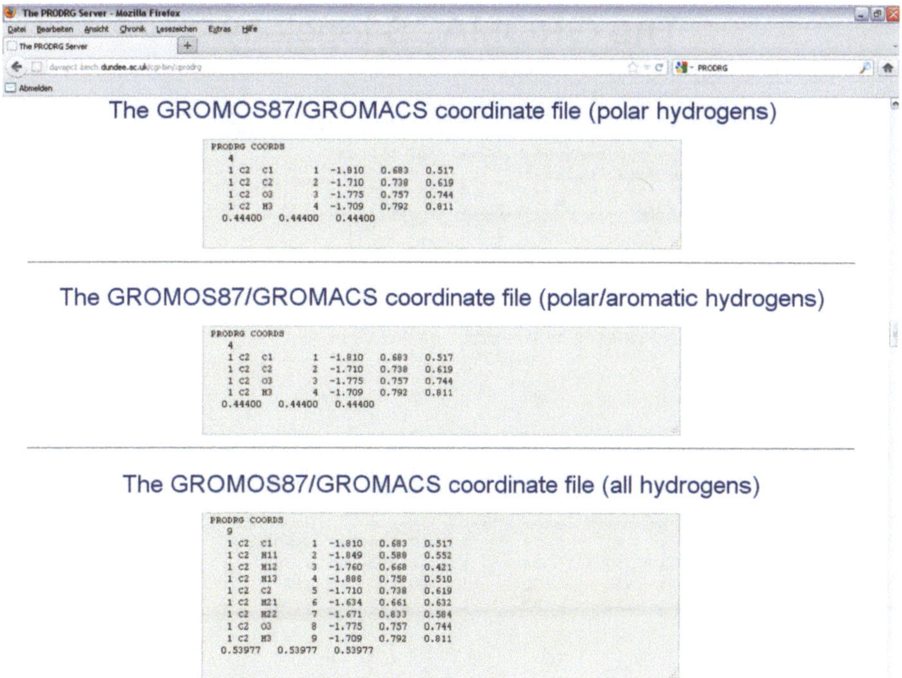

Fig. 4.5 Three different GROMOS87/GROMACS coordinate files as output of the PRODRUG run

to be the same as in the topology file. Thus, copy all lines within the box titled "polar/aromatic hydrogens" and save them in a file named `ethanol.gro`.

Now, scroll down to the section "The GROMACS topology" (Fig. 4.6), copy the contents and save in a file named `ethanol.itp`.

Now, you have all data for performing simulations including ethanol with GROMACS.

In the following box, a summary of all steps for generating a GROMACS coordinate- and topology-file is given:

The GROMACS topology

```
;
;
;       This file was generated by PRODRG version AA081006.0504
;       PRODRG written/copyrighted by Daan van Aalten
;       and Alexander Schuettelkopf
;
;       Questions/comments to dava@davapc1.bioch.dundee.ac.uk
;
;       When using this software in a publication, cite:
;       A. W. Schuettelkopf and D. M. F. van Aalten (2004).
;       PRODRG - a tool for high-throughput crystallography
;       of protein-ligand complexes.
;       Acta Crystallogr. D60, 1355--1363.
;
;

[ moleculetype ]
; Name nrexcl
LIG      3

[ atoms ]
;   nr      type  resnr resid  atom  cgnr   charge     mass
     1       CH3     1  LIG      C4     1    0.074  15.0350
     2       CH2     1  LIG      C1     1    0.091  14.0270
     3        OA     1  LIG      O2     1   -0.202  15.9994
     4         H     1  LIG      H2     1    0.037   1.0080

[ bonds ]
; ai  aj  fu    c0, c1, ...
   2   1   2    0.153   7150000.0    0.153   7150000.0 ;    C1   C4
   2   3   2    0.143   8180000.0    0.143   8180000.0 ;    C1   O2
   3   4   2    0.100  15700000.0    0.100  15700000.0 ;    O2   H2

[ pairs ]
; ai  aj  fu    c0, c1, ...
   1   4   1                                          ;    C4   H2

[ angles ]
; ai  aj  ak  fu    c0, c1, ...
   1   2   3   2    109.5       520.0     109.5       520.0 ;    C4   C1   O2
   2   3   4   2    109.5       450.0     109.5       450.0 ;    C1   O2   H2

[ dihedrals ]
; ai  aj  ak  al  fu    c0, c1, m, ...
   1   2   3   4   1       0.0     1.3 3       0.0     1.3 3 ; dih   C4   C1   O2
H2
```

Fig. 4.6 GROMACS-topology-file of ethanol, calculated by the PRODRG server

- ☐ Construct your ligand with an appropriate editor as 3D-structure
- ☐ Check your molecule, if e.g. the configuration of chiral atoms is correct
- ☐ Minimize your molecule, if possible
- ☐ Save the minimized molecule as pdb-file
- ☐ Start the PRODRG-Server
- ☐ If you do not have a token to work with the PRODRG-Server, please, fill in your E-Mail in the corresponding field and used the "Send" button. Be aware, that it may take some time, before you get your token via E-Mail
- ☐ If you have the token, please copy it from your E-Mail into the appropriate field. Now, you can start working with PRODRG.
- ☐ Open your pdb-file in an appropriate editor
- ☐ Copy and paste the whole pdb-file into the corresponding field of the PRODRG-server
- ☐ Choose "Yes" or "No" in the field chirality (depends on your molecule)
- ☐ Always choose "full charges" in the field charges
- ☐ Choose "Yes" in the field EM (energy minimization)
- ☐ Now, start your PRODRG-Job. Please be aware, that the calculations may take a while and do not close your browser
- ☐ Copy your GROMACS coordinates with polar/aromatic hydrogens and save them as `gro`-file
- ☐ Copy your GROMACS topology and save it as `itp`-file
- ☐ Load your gromacs-coordinate file into an editor for visualization of molecules and verify the structure

Next, we present another example, dealing with the ligand dobutamin (Fig. 4.7).

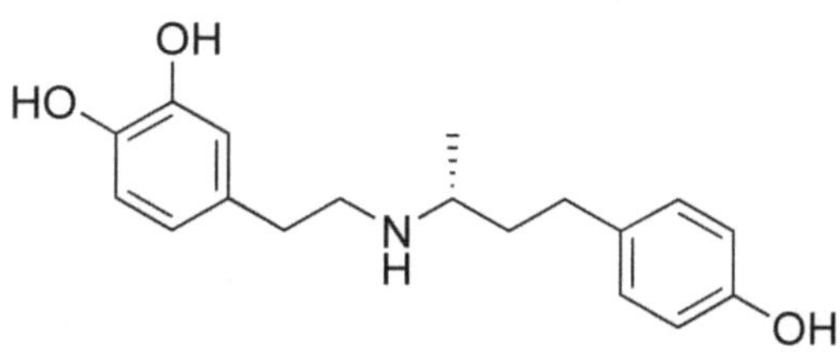

Fig. 4.7 Structure of dobutamine (only R enantiomer shown), cocrystallized with the turkey β_1 adrenergic receptor in the crystal structure 2Y00. (Warne et al. 2011)

Dobutamine is a β_1-sympathomimetic drug. It is cocrystallized with the turkey β_1 adrenergic receptor in the crystal structure 2Y00 (Warne et al. 2011).

Exercise Please upload the dobutamine in its neutral form, as pointed out in Fig. 4.7 to the PRODRG-server and create the appropriate files, as mentioned above. If you have done so, you should have a closer look onto the output of the PRODRG-server (Fig. 4.8). You can see, that now the dobutamine is positively charged, since there is an additional hydrogen at the amino moiety. Remember, you uploaded dobutamine in its neutral form. Be aware, that this is typical for the PRODRG-server: molecules with a basic or acid moiety are calculated in its charged form.

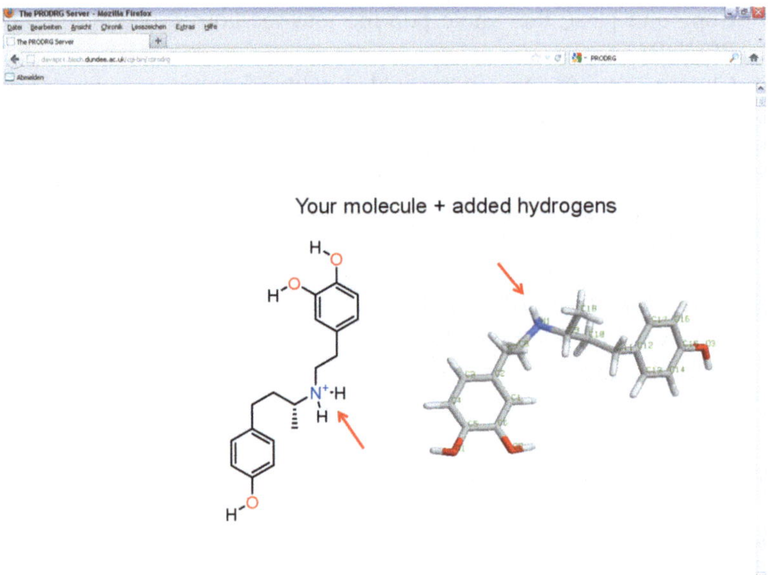

Fig. 4.8 Output of the PRODRG-Server with regard to dobutamine

The action, that molecules with a basic or acid moiety are calculated in its charged form, is based on the behaviour of carboxylic acid or amines in water at pH values about 7. Under these conditions, carboxylic moieties for example are deprotonated, and amino moieties are protonated.

Chapter 5
Lipids

Lipid membranes separate two compartimentes from each other: they separate a cell from the surrounding, or they separate the cytoplasm of cells into organelles. These membranes consist of two layers of lipid, the so called lipid bilayer. The lipid bilayer is a planar, two dimensional fluid.

A large number of proteins belong to the class of membrane proteins. Membrane proteins can be divided into two groups: First, peripheral membrane proteins, which are located on the surface of the lipid bilayer and second, the integral membrane proteins. It is typical for integral membrane proteins, that they are embedded into the phospholipid bilayer. GPCRs belong to the membrane proteins and are also called 7TM receptors, since they consist of 7 transmembrane domains, which cross the lipid bilayer. These transmembrane domains are connected by sections with some few up to some hundreds of amino acids, which are located in the aequeous extra- and intracellular sides of the lipid bilayer.

Within the first molecular modelling studies of GPCRs, the GPCRs were modelled in the gas phase. This was a very rigorous approximation, because, the amino acid side chains, pointing outwards of the receptor, were not in contact with the native surrounding. This could lead to incorrect amino acid side chain conformations, or to artificial interactions between polar or charged amino acids. Additionally, if molecular dynamic simulations were performed of a GPCR in the gas phase, the secondary and tertiary structure of the receptor was not stable. In order to achieve stability, constraints had to be put onto the backbone of the protein. Thus, conformational changes with regard to the whole receptor could not be observed. But with the development of more efficient computers, it was possible to simulate GPCRs in their natural surrounding, like lipid bilayer including intra- and extracellular water. Meanwhile, it is widespread established, to model a GPCR in its natural surrounding.

5.1 Structure of Lipids

Lipids can be divided into several groups, the phosphoglycerides, sterols, sphingolipids, triglycerides and glycolipids. Membrane bilayers are mainly constituted by phosphoglycerides. A schematic representation of phophoglycerides is given in Fig. 5.1.

A. Strasser, H.-J. Wittmann, *Modelling of GPCRs*,
DOI 10.1007/978-94-007-4596-4_5,

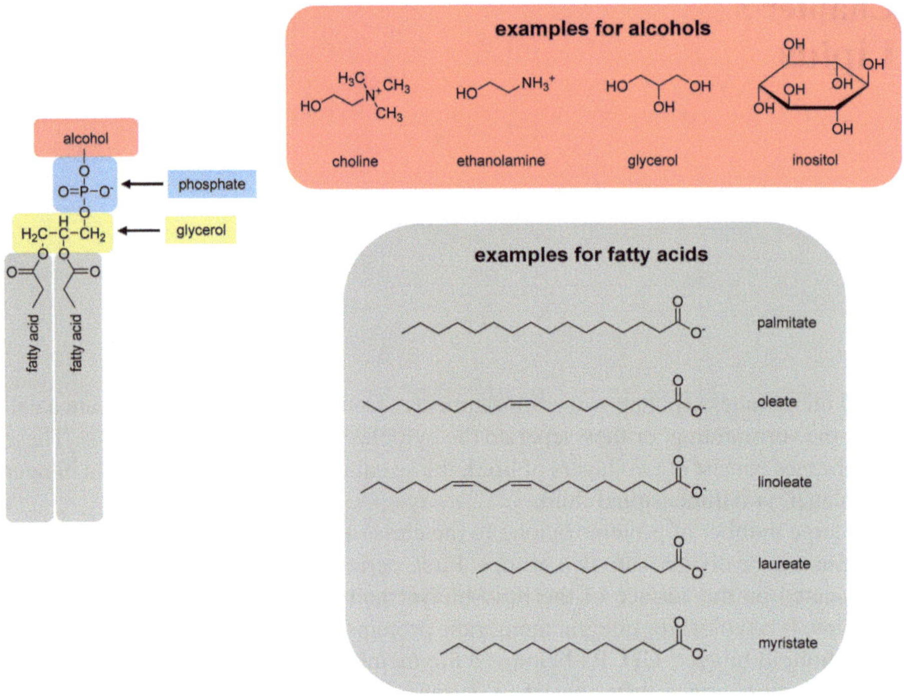

Fig. 5.1 Structure of phosphoglycerides

The phosphoglycerides are established by one glycerol. Two long-chain fatty acids are esterified to the carbons C1 and C2 of the glycerol. The fatty acids are carboxylic acids with about 12–20 carbon atoms. A phosphoric acid is esterified to C3 of the glycerol and an alcohol to the phosphate. Due to their chemical structure, phosphoglycerides are amphiphilic. The head groups are hydrophilic, whereas the long fatty acids show hydrophobic properties. In biological systems, a large variety of phosphoglycerids is found, since there is a high variability with regard to the alcoholic group and the fatty acids.

The name of the phosphoglycerides is based on the alcoholic head groups:

- Phosphatidic acid, **PA** (no head group), i.e. PO**PA**
- Phosphatidylcholine, PC, i.e. PO**PC**
- Phosphatidylethanolamine, **PE**, i.e. PO**PE**
- Phosphatidylglycerol, PG, i.e. PO**PG**
- Phosphatidylinositol, PI
- Phosphatidylserine, PS, i.e. PO**PS**

The PO in the lipids mentioned above, is the abbreviation for 1-palmitoyl-2-oleol. For MD simulations, GPCRs are mainly embedded into POPC lipid bilayers (Ivanov et al. 2005; Filizola et al. 2006; Henin et al. 2006; Strasser et al. 2007). The structure of POPC is presented in Fig. 5.2. However, other lipid models, like DOPC (dioleoylphosphatidylcholine) are used (Goetz et al. 2011).

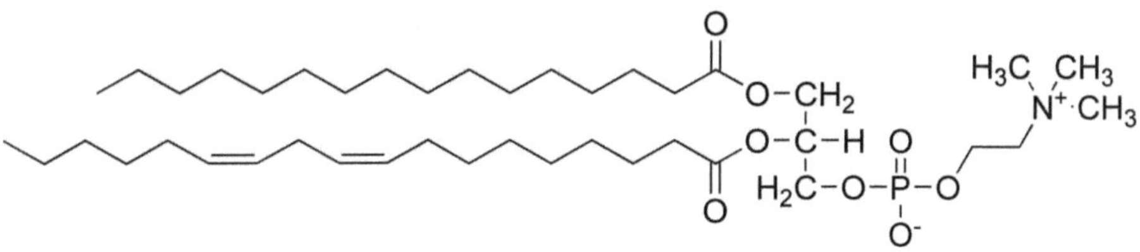

Fig. 5.2 Structure of POPC (1-palmitoyl-2-oleoylphosphatidylcholine)

Sterols are another important class of membrane lipids. One of the most prominent is the cholesterol (Fig. 5.3). The cholesterol scaffold contains four condensed rings leading to a distinct rigidity. This structure is hydrophobic and thus it is able to insert into the hydrophobic inner layer of the lipid bilayer. The polar hydroxyl moiety is located at the surface of the lipid bilayer.

Fig. 5.3 Structure of cholesterol

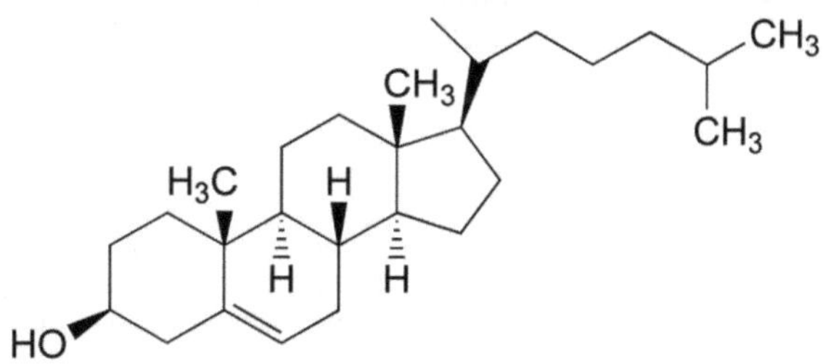

Cholesterol is sometimes found cocrystallized in combination with crystal structures of GPCRs. For example, a cholesterol specific binding site was identified for the human β_2 adrenergic receptor within the crystal structure 3D4S (Hanson et al. 2008).

5.2 Structure of the Phospholipid Bilayer

In Fig. 5.4, a site model of a lipid bilayer is presented. The hydrophobic chains point inside the lipid bilayer, whereas the polar head groups are facing towards the surrounding water.

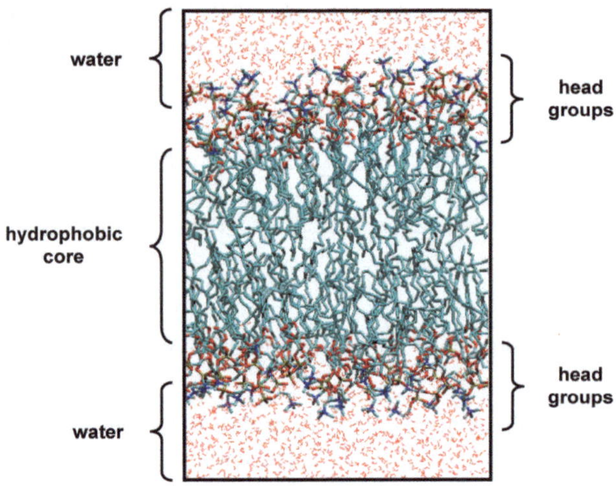

Fig. 5.4 Model of a lipid bilayer with water on both sides

5.3 Lipid Bilayer Models Used in Molecular Modelling

Several lipid models were constructed for use in molecular modelling. Some of them are summarized in Table 5.1.

Table 5.1 Summary of some lipids, often used in molecular modelling studies

DPPC	Dipalmitoylphosphatidylcholine
DMPC	Dimyristoylphosphatidylcholine
DOPC	Dioleoylphosphatidylcholine
POPC	1-palmitoyl-2-oleoylphosphatidylcholine
POPE	1-palmitoyl-2-oleoylphosphatidylethanolamine
PLPC	Palmitoyllinoleylphosphatidylcholine

5.4 Internet Sources for Lipid Bilayer Models

In the internet, there are some sources which give a more detailed information with regard to lipid bilayers, including simulation parameters for GROMACS. At some sites in internet, equilibrated lipid bilayer models can be obtained via free download. A summary of the most important internet resources with regard to lipids is given in Table 5.2.

Table 5.2 Most important internet resources with regard to lipds

URL
http://lipidbook.bioch.ox.ac.uk
http://moose.bio.ucalgary.ca/index.php?page=Structures_and_Topologies
http://www.lrz-muenchen.de/~heller/membrane/membrane.html
http://www.scmbb.ulb.ac.be/Users/lensink/lipid/

A very comfortable site is lipidbook (http://lipidbook.bioch.ox.ac.uk) (Domanski et al. 2010). The aim of the lipidbook is "a public repository for force field parameters with special emphasis on lipids" (http://lipidbook.bioch.ox.ac.uk) (Fig. 5.5). Here, you can individually select the force-field, the parameter notation for distinct software and the kind of lipid (Fig. 5.6).

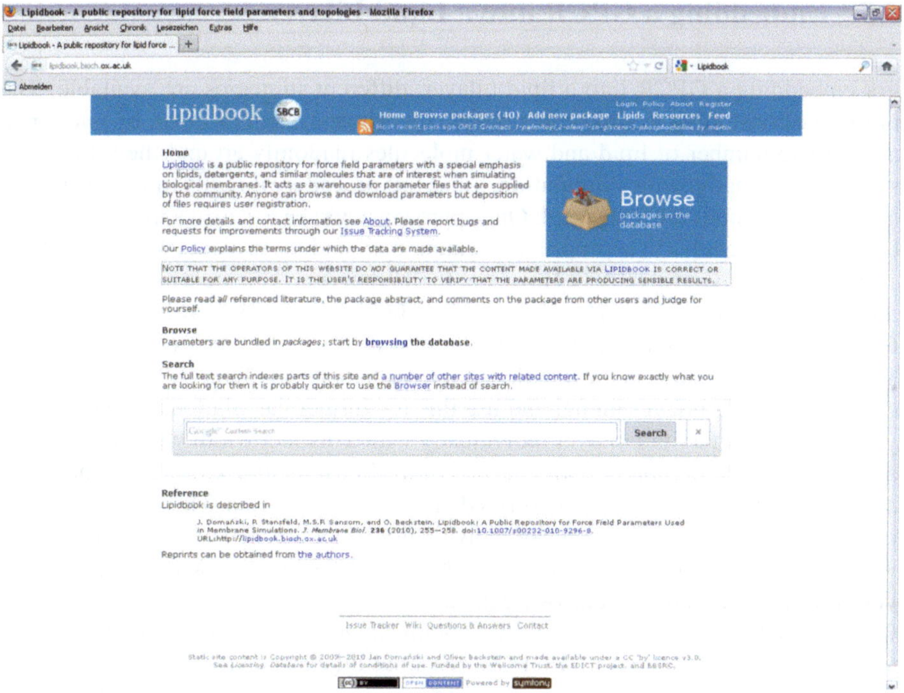

Fig. 5.5 Starting page of lipidbook (http://lipidbook.bioch.ox.ac.uk). (Domanski et al. 2010)

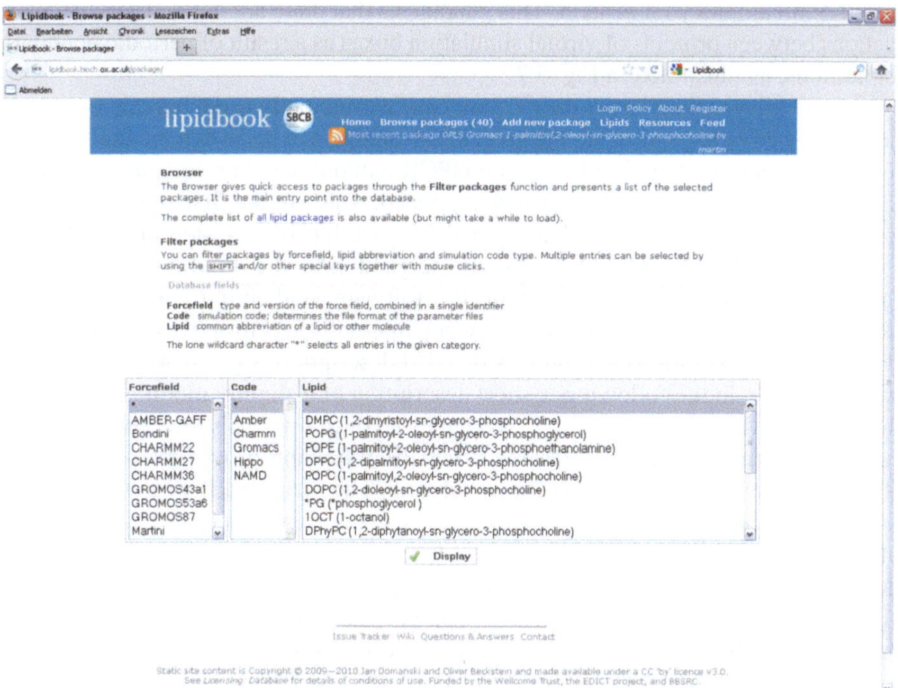

Fig. 5.6 Browser in lipidbook (http://lipidbook.bioch.ox.ac.uk). (Domanski et al. 2010)

5.5 Embedding a GPCR into a Lipid Bilayer

For embedding a GPCR into a lipid bilayer, different strategies are available. The most time consuming would be to simulate the whole system *de novo*, by putting an appropriate number of lipid and water molecules randomly around the GPCR and start a molecular dynamics simulation. Since this procedure is really time consuming, alternative methods are suggested: One approach could be to set an appropriate number of lipid molecules in appropriate orientation around the protein (Woolf and Roux 1996; Belohorcova et al. 1997). However, for this strategy, you must have access to an appropriate software, or you have to establish the software by yourself. Alternatively, for setting up your simulation box, you can start with already prepared lipid bilayers. Therefore, you can look at the mentioned internet resources (Table 5.2), download an equilibrated lipid bilayer model and use this for further calculations. Alternatively, you can construct a lipid bilayer individually with a distinct width with an appropriate software. One suitable software is `vmd` (http://www.ks.uiuc.edu/Research/vmd/), combined with some scripts, as described in more detail later on. The great advantage of the latter strategy is that you can individually adopt the size of your lipid bilayer with regard to the size of the GPCR or the GPCR-Gαβγ-complex. In this context you have to take into account two considerations: What do you want to simulate: Only a GPCR or a whole GPCR-Gαβγ-complex. Due to the larger size of a GPCR-Gαβγ-complex, compared to a GPCR, the lipid bilayer has to be large in case of a GPCR-Gαβγ-complex. However, in both cases, the lipid bilayer must be large enough in order to guarantee that the GPCR or GPCR-Gαβγ-complex is embedded well. This means, you should have a lipid bilayer with a width of optimally 1.0–1.5 nm around your protein. This guarantees that there are not undesirable interactions between proteins of virtual simulation boxes as a result of periodic boundary conditions, as illustrated in Fig. 5.7.

A lipid bilayer shell larger than 1.5 nm can be principally used, but this would not lead to any advantage, instead, the great disadvantage will be an exponential increase in simulation time. For simulation of a GPCR without the G protein, a width of the lipid bilayer of about 9–10 nm is recommended. Thus, in the first step you have to generate your lipid bilayer with an appropriate width (Fig. 5.8, step 1).

Subsequently, the GPCR has to be aligned into the lipid bilayer. A very good description in combination with the software vmd (http://www.ks.uiuc.edu/Research/vmd/) is found at the following internet site: http://www.ks.uiuc.edu/Research/vmd/plugins/membrane/. A detailed description is given at the mentioned site. However, in the following, a short description of a slightly modified procedure using the script `combine.tcl`, available at (http://www.ks.uiuc.edu/Research/vmd/plugins/membrane/) is presented. For the following procedure you need the shell script `vmd2gro`, which is shown later on. The script `vmd2gro` was tested in combination with `vmd 1.8.7`. Be aware, that in the presented version of `vmd2gro` the POPC molecules in vmd-notation are transferd into the POPC-notation used by Moose (http://moose.bio.ucalgary.ca/index.php?page=Structures_and_Topologies). Thus, for further use with GROMACS, you need the files `lipid.itp` and `popc.itp`. Both are available at http://moose.bio.ucalgary.ca/index.php?page=

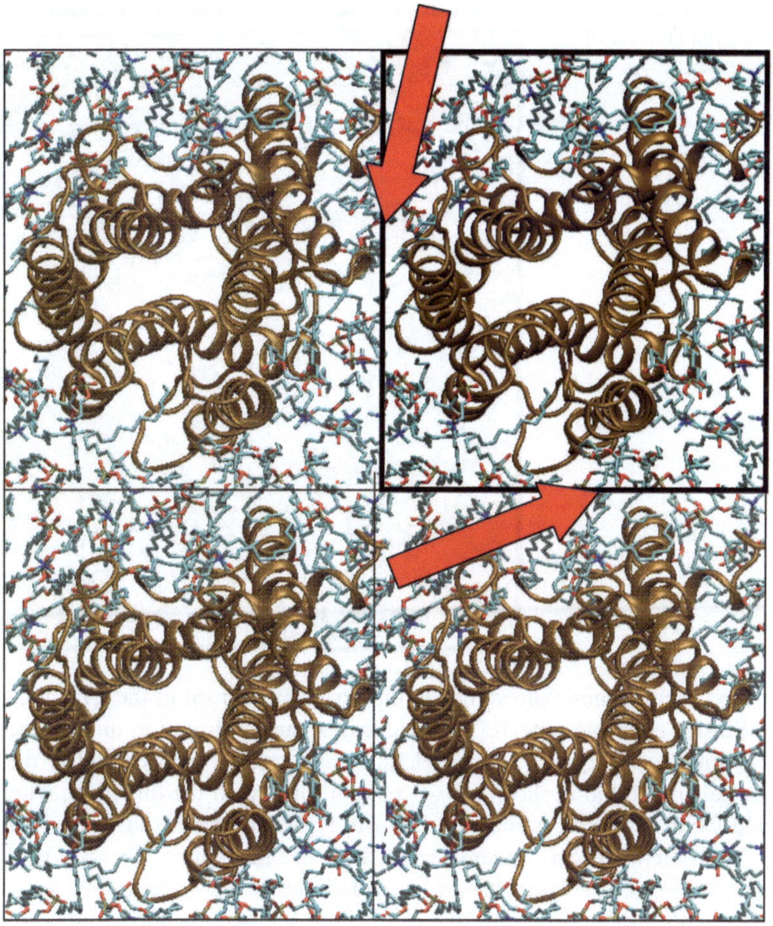

Fig. 5.7 Artificial close contacts of the protein in the simulation box with the proteins in the neighboured virtual simulation boxes due to periodic boundary conditions

Structures_and_Topologies and can also be found in the appendix (POPC Parameters). In the script `vmd2gro` the tcl-script `combine.tcl`, developed by Balabin and published at http://www.ks.uiuc.edu/Research/vmd/plugins/membrane/ was used (see lines 192–276 in `vmd2gro` shown below). Before starting, you need your protein as a **correct** pdb-file, which can be used as an input file for the GROMACS command `pdb2gmx` without special options. In this example, this file is named `protein.pdb`. In the next step, you can start `vmd` to generate the lipid bilayer. Therefore, choose in the main menu `Extentions` → `Modelling` → `Membrane Builder`. There, choose `POPC` as lipid, because the script `vmd2gro` only considers `POPC` in the version presented here. Define the length of the lipid bilayer in x- and y-direction. Be aware to define the values in Å. In the actual version of the script `vmd2gro` a box size of about (10 × 10 × 10) nm is defined (see line 188).

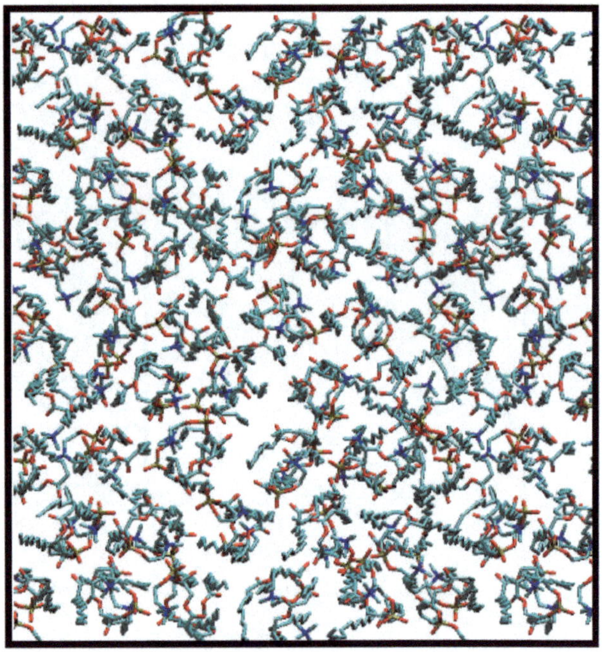

Fig. 5.8 Step 1: Generate or download a lipid bilayer of appropriate size

However you can change the values in the script of later on in the gro-file, if necessary. So, for example type 100 for width in x- and y-direction in the appropriate field of vmd. As "ouput prefix" type membrane. Subsequently, the membrane is generated and is shown in vmd. Additionally, in the directory, where vmd is started from, two files, named membrane.pdb and membrane.psf are generated. The extension psf means "protein structure file". In the next step, the file protein.pdb has to be loaded via File → New Molecule. To simplify the alignment of the protein in the lipid bilayer, use the menu Graphics → Representations → "Coloring Method → Color ID → yellow" and "Drawing Method → New Cartoon". Now, the coordinates of the protein have to be changed via Mouse → Move → Molecule. Be careful and move **ONLY** the protein and **NOT** the membrane. Moving the membrane in the "move" mode would result in a failure of the alignment! In the "move" mode, use the left mouse button for translation, the shift-button and the left mouse button for rotation around the z-axis and the shift-button in combination with the middle mouse button to rotate around the axis vertical to the screen. Therefore, click directly onto the protein with the mouse cursor. To leave the "move" mode, type "r" or use the menu button Mouse → Rotate. In the "rotate" mode, you can rotate the whole system (membrane and protein) as appropriate without changing any coordinates. In the next step, you have to use the "*move*" and "*rotate*" mode alternately to align the protein into the lipid bilayer. This procedure should be performed very carefully with regard to the placement of the GPCR in the membrane. Additionally, this procedure needs some practice and may take some time. If the alignment procedure is finished, the protein with the new coordinates has to be saved as `pdb`- and `psf`-file in the following manner: Use the menu buttons `Extensions` → `Modelling` → `Automatic PSF Builder` and a new window, opens: Step 1: In the field "`Output basename`", we write `protein_autopsf` for example and click onto the button "`Load input files`". Step 2: Choose "`Everything`" and click onto "`Guess and split chains using current selections`". Step 3: Click onto "`Create chains`". Step 4: Click onto "`Apply patches and finish PSF/PDB`". An additional window opens, there, click "`OK`" and finish by clicking onto the button "`Reset Autopsf`". If you look into the directory, where `vmd` is started from, three new files `protein_autopsf.log`, `protein_autopsf.pdb` and `protein_autopsf.psf` are generated by the procedure, mentioned above. Now, `vmd` can be closed and the shell script `vmd2gro` can be startet:

```
> vmd2gro ↵
```

Subsequently, you have to define some basenames of files:

```
> Basename of membrane file: membrane ↵
> Basename of aligned protein file: protein_autopsf ↵
> Basename of output file (protein+membrane): temp ↵
```

The first two basenames have to be the same as used in the alignment procedure mentioned above. The third can have any basename, since these will be temporary files, which will be deleted automatically.

After that, the script performs some calculations and then it stops in order to ask you, if protonation states of amino acids should be changed. Here answer "*no*" by typing a "`n`". Subsequently, the command `pdb2gmx` is called within `vmd2gro` and you are asked to choose an appropriate force-field. For example, type "`4`". Now, `vmd2gro` performs some time-consuming calculations, like generation of a topology file. After some minutes, `vmd2gro` should have finished. Now, you should have some new files in your current working directory: `membrane.gro`, `protein_autopsf.gro`, `protein_autopsf.gro`, `protein_autopsf.top` and `posre.itp`. The files `membrane.gro` and `protein_autopsf.gro` contain the coordinates, relevant for the further steps. In `membrane.gro`, the POPC-lipid-bilayer with a hole and in an appropriate site-notation is given. Be aware, that the POPC in this notation can only be used with the parameters available in internet (http://moose.bio. ucalgary.ca/index.php?page=Structures_and_Topologies) and shown explicitly in the appendix (POPC Parameters). The file `protein_autopsf.gro` contains the coordinates of the aligned protein and the file `protein_autopsf.top` is the corresponding topology-file. The coordinates of both `gro`-files, `membrane.gro` and `protein_autopsf.gro` can be combined within one gro-file, containing now the protein and lipids. To do so, one can use the following LINUX-command-sequence:

```
> set nr_prot  =`wc -l protein_autopsf.gro | cut -d' ' -f1` ↵
> set nr_mem = `wc -l membrane.gro | cut -d' ' -f1` ↵
> @ all_sites = ${nr_prot} + ${nr_mem} - 6  ↵
> echo "Protein in lipid bilayer" > prot_lip.gro ↵
> echo "$all_sites" >> prot_lip.gro ↵
> tail -n +3 protein_autopsf.gro | head -n -1 >> prot_lip.gro ↵
> tail -n +3 membrane.gro >> protein_lip.gro ↵
```

Now, you should have your protein and the lipid bilayer in the file `protein_lip.gro`. Of course, you can do the analogue manipulations manually with an editor. In `protein_lip.gro`, the lipid sites start again with number 1. To obtain a subsequent numbering and centring the structure in the simulation box, use the GROMACS command `editconf`:

```
> editconf –f protein_lip.gro -c –o system.gro ↵
```

The user of `vmd2gro` and the tcl-script `combine.tcl` developed by Balabin and published at http://www.ks.uiuc.edu/Research/vmd/plugins/membrane/ (see lines 192–276 in `vmd2gro` shown below) should add the absolute path for `top_all27_prot_lipid.inp` in line 207. The script `vmd2gro` is shown in the following:

```
1 #!/bin/tcsh
2
3 # vmd2gro: tcsh script to convert vmd format to gro format
4 # for a lipid membrane/protein complex
5
6 # Detect and remove collisions between protein and membrane using a Tcl script for VMD
7 # script statements follow a line beginning with the the pattern "# START_TCL"
8
9 # Variables containing file names: mem, prot, out
10 # Tcl scriptfile "out.tcl" will be created on the fly
11
12 echo -n "Basename of membrane file: "
13 set mem = "$<"
14
15 if (! -e "${mem}.pdb") then
16   echo "Missing pdb file: ${mem}.pdb"
17   exit 1
18 else if (! -e "${mem}.psf") then
19   echo "Missing psf file: ${mem}.psf"
20   exit 1
```

```
endif

echo -n "Basename of aligned protein file: "
set prot = "$<"

if (! -e "${prot}.pdb") then
  echo "Missing pdb file: ${prot}.pdb"
  exit 1
else if (! -e "${prot}.psf") then
  echo "Missing psf file: ${prot}.psf"
  exit 1
endif

echo -n "Basename of output file (protein+membrane): "
set out = "$<"

if (-e "${out}.pdb") then
  echo "File ${out}.pdb exists!"
  echo "Rename existing file or choose new file name and start again"
  exit 1
else if (-e "${out}.psf") then
echo "File ${out}.psf exists!"
echo "Rename existing file or choose new file name and start again"
  exit 1
endif

# Create and start tcl script: out.tcl

if (-e "${out}.tcl") then
echo "Tcl script file ${out}.tcl exists! Remove or rename it!"
  exit 1
endif

# Substitutions for MEM, PROT and OUT in Tcl script part

set begin_tcl = `grep -n "^# START_TCL" $0|cut -d ':' -f1`
echo $begin_tcl

tail -n +$begin_tcl $0|sed -e "s/MEM/$mem/"
```

```
-e "s/PROT/$prot/" -e "s/OUT/$out/"
> "${out}.tcl"
if ("$status") then
echo "Error creating Tcl script file
${out}.tcl! Terminating!"
exit 1
endif
# Starting Tcl script from working directory!
chmod u+x "${out}.tcl"
vmd -e ${out}.tcl -dispdev text
if ("$status") then
  echo "${out}.tcl failed! Terminating
calculations!"
exit 1
endif
echo "Output files ${out}.pdb and ${out}.psf
successfully created!"
echo "Going to convert pdb format to gro format"
# Removing obsolete files: pdb and psf files
of membrane and protein; tcl script file;
output psf file
rm "${mem}".{pdb, psf} "${prot}".{pdb, psf}
"${out}".tcl "${out}".psf
# Extracting protein and membrane structures
from out (*.pdb) file, deleting all water
molecules
grep 'POPC' "${out}.pdb" > "${mem}.pdb"
grep -v 'TIP\|POPC' "${out}.pdb"|grep 'ATOM'|
 sed -e 's/HSD/HIS/' > "${prot}.pdb"
rm "${out}.pdb"
# Converting pdb file for protein and possibly
change protonation state
```

```

echo "*********************************************************************************************"
echo -n "Do you want to modify the protonation state of amino acid ARG, ASP, GLU, HIS or LYS? (y/n): "
echo "*********************************************************************************************"

set answer = "$<"
set answer = `echo $answer|tr 'y' 'Y'`

set ARG = ""
set ASP = ""
set GLU = ""
set HIS = ""
set LYS = ""

if ("$answer" == "Y") then
  echo -n "Change protonation state of ARG? (y/n): "
set h = "$<"
set h = `echo $h|tr 'y' 'Y'`
if ("$h" == "Y") then
  set ARG = "-arg"
endif

echo -n "Change protonation state of ASP? (y/n): "
set h = "$<"
set h = `echo $h|tr 'y' 'Y'`
if ("$h" == "Y") then
  set ASP = "-asp"
endif

echo -n "Change protonation state of GLU? (y/n): "
set h = "$<"
set h = `echo $h|tr 'y' 'Y'`
if ("$h" == "Y") then
  set GLU = "-glu"
endif

echo -n "Change protonation state of HIS? (y/n): "
set h = "$<"
set h = `echo $h|tr 'y' 'Y'`
if ("$h" == "Y") then
```

```
  set HIS = "-his"
endif
echo -n "Change protonation state of LYS? (y/n): "
set h = "$<"
set h = `echo $h|tr 'y' 'Y'`
if ("$h"=="Y") then
  set LYS = "-lys"
endif
endif
pdb2gmx -f "${prot}.pdb" -o "${prot}.gro" -p "${prot}.top" -ignh $ARG $ASP $GLU $HIS $LYS
rm "${prot}.pdb"
# Converting pdb file for membrane ******************************************
setenv LC_NUMERIC '.'
# Initializations -map VMD atomic numbers to GRO atomic numbers (index)
set map = (4 3 5 1 2 17 23 20 21 22 24 25 28 30 31 32 33 45 48 51 54 57 60 63 65 67 70 73 76 79 82 36 39 40 41 42 92 95 98 101 104 107 110 113 116 119 122 125 128 131 85 88)
set gro_label = (C1 C2 C3 N4 C5 C6 O7 P8 O9 O10 O11 C12 C13 O14 C15 O16 C17 C18 C19 C20 C21 C22 C23 C24 C25 C26 C27 C28 C29 C30 C31 C32 O33 C34 O35 C36 C37 C38 C39 C40 C41 C42 C43 C44 C45 C46 C47 C48 C49 C50 CA1 CA2)
set n_gro = $#map # number of sites per lipid molecule in gro notation
set n_pdb = 134 # number of atoms per lipid molecule in VMD notation
set gro_file = "${mem}.gro"
# Caluclations and mapping
```

```
@ units = 'wc -l "${mem}.pdb" | cut -d ' ' -f1' / $n_pdb # number of lipid molecules
@ number_of_atoms = $units * $n_gro

echo "Lipid membrane" >> $gro_file
echo "$number_of_atoms" >> $gro_file

set unit = 1
set atom_no = 1
while ("$unit" <= "$units")
  set n = 1
  @ i1 = ($unit - 1) * $n_pdb
  while ("$n" <= "$n_gro")
    @ pdb_line = $i1 + $map[$n]

    gawk -v u=$unit -v line=$pdb_line -v atom=$atom_no -v label=$gro_label[$n] 'NR==line {printf("%5i%3s%7s%5i%8.3f%8.3f%8.3f\n", u,"POP",label, atom,$6/10.0,$7/10.0,$8/10.0)}' "${mem}.pdb" >> $gro_file

    @ n++
   @ atom_no++
  end

  @ unit++
end

echo " 10.00 10.00 10.00" >> $gro_file

exit 0

#*************************************************
# START_TCL script part; do not edit or delete this label! ***
# *** Following tcl commands for VMD ***
# embed (parts of) protein into a membrane
# Ilya Balabin (ilya@ks.uiuc.edu), 2002-2003
#
# You need: a) membrane structure (membrane.psf/pdb);
# b) properly oriented and aligned to the membrane
# protein structure (protein.psf/pdb)


```

```
# set echo on for debugging
echo on

# need psfgen module and topology
package require psfgen
topology top_all27_prot_lipid.inp

# load structures
resetpsf
readpsf MEM.psf
coordpdb MEM.pdb
#readpsf protein.psf
readpsf PROT.psf
#coordpdb protein_aligned.pdb
coordpdb PROT.pdb

# can delete some protein segments; list them in brackets on next line
set pseg2del { }
foreach seg $pseg2del {
  delatom $seg
}

# write temporary structure
set temp "temp"
writepsf $temp.psf
writepdb $temp.pdb

# reload full structure (do NOT resetpsf!)
mol load psf $temp.psf pdb $temp.pdb

# select and delete lipids that overlap protein:
# any atom to any atom distance under 0.8A
# (alternative: heavy atom to heavy atom distance under 1.3A)
set sellip [atomselect top "resname POPC"]
set lseglist [lsort -unique [$sellip get segid]]
foreach lseg $lseglist {
# find lipid backbone atoms
set selover [atomselect top "segid $lseg and within 0.8 of protein"]
# delete these residues
set resover [lsort -unique [$selover get resid]]
foreach res $resover {
  delatom $lseg $res
```

```
}
}
# delete lipids that stick into gaps in protein
foreach res { } {delatom $LIP1 $res}
foreach res { } {delatom $LIP2 $res}
# delete lipids that fall out of the PBC box
# the following numbers are for example only; yours are different!
set xmin -55
set xmax 41
set ymin -51
set ymax 34
foreach lseg {"LIP1" "LIP2"} {
# find lipid backbone atoms
set selover [atomselect top "segid $lseg and (x< $xmin or x> $xmax or y< $ymin or y> $ymax)"]
# delete these residues
set resover [lsort -unique [$selover get resid]]
foreach res $resover {
  delatom $lseg $res
}
}
# write full structure
writepsf OUT.psf
writepdb OUT.pdb
# clean up
file delete $temp.psf
file delete $temp.pdb
# non-interactive script
quit
```

In Fig. 5.9, the membrane with a hole, created as described above, is shown. The file `membrane.gro` should look similar, if loaded into `vmd`.

After establishing an appropriate hole in the lipid bilayer and putting the GPCR into the hole, you should receive a system as shown in Fig. 5.10. The file `protein_lip.gro`, created above, should look similar. As you can see in the figure, there is a significant gap between the lipid bilayer and the GPCR.

Now, the system consists of the lipid bilayer and the GPCR. Using the GROMACS commands `grompp` and `mdrun`, the system can be minimized (see Chap. 6). Thus,

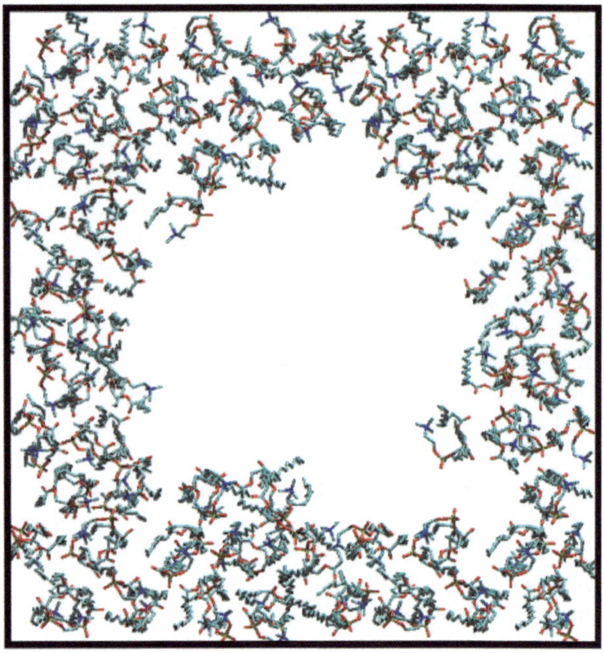

Fig. 5.9 Step 2: Generate a hole of appropriate size for the GPCR in the lipid bilayer

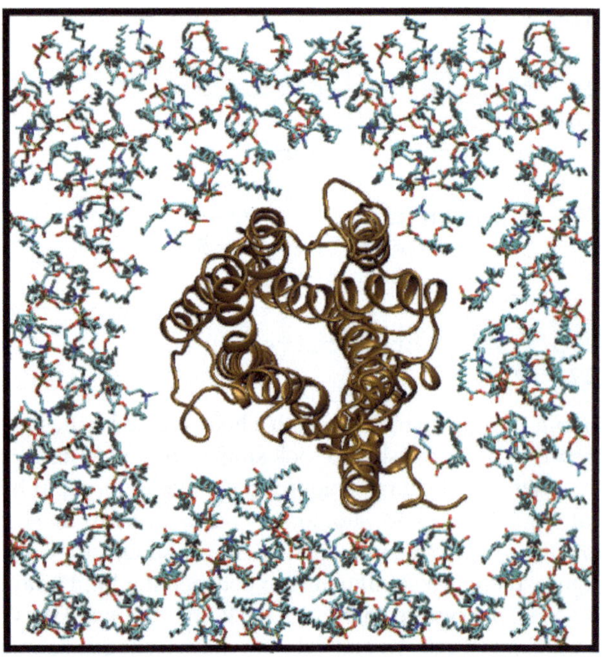

Fig. 5.10 Step 3: Placement of the GPCR or GPCR-G-protein-complex in the lipid bilayer

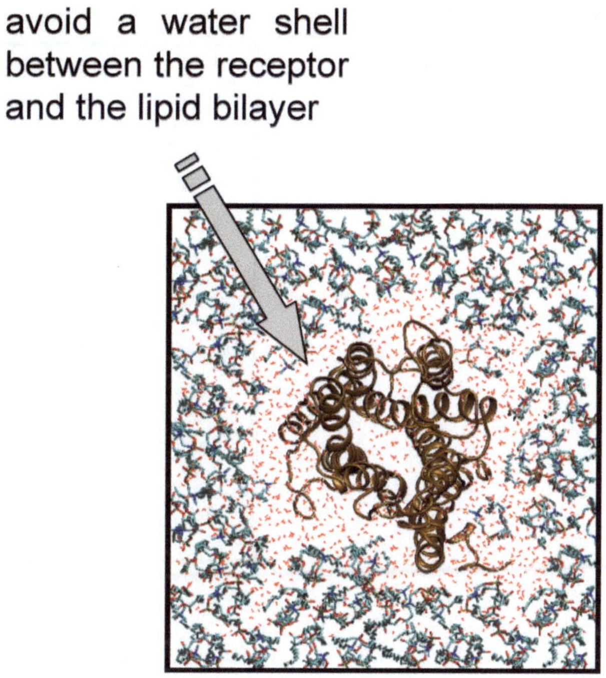

Fig. 5.11 An artificial water shell between the GPCR and the lipid bilayer, as consequence of wrong system setup

one might think, that the system has to be solvated in the next step. Doing so would lead to an artificial system, as pointed out in Fig. 5.11.

Due to the gap between the GPCR and the lipid bilayer, a large number of water molecules would be put into this gap during the solvation of the system. This water between the GPCR and the lipid bilayer is artificial and may lead to problems during the simulation or to artefacts, because the hydrophobic transmembrane domains of the receptor and the hydrophobic fatty acid side chains of the lipids are in contact to the hydrophilic water. Thus, both, the hydrophobic side chains and lipids might obtain energetically more favoured conformations without contact to the hydrophilic water. This may lead to instabilities of the receptor during simulation. However, some 10 water molecules all in all between lipid and receptor should not lead to problems during the simulation. They can be removed, but in most cases, they move into the extra- or intracellular water during the simulation. In order to avoid scenarios, as illustrated in Fig. 5.11, the lipid bilayer should be equilibrated around the GPCR (Fig. 5.12) before solvating the system. Therefore, different simulation protocols can be used. However, positions constraints have to be put at least onto the protein in order to avoid any conformational change of the protein during the lipid-equilibration process. In order to obtain an equilibration of the lipids in the xy-plain, slight position constraints might be put onto the z-coordinates of the lipids. In general, the modeller is encouraged to perform some different equilibration protocols in order to obtain an optimal structure. After this equilibration step, the lipid bilayer is fitted well to the GPCR and the gap between the GPCR and the lipid bilayer is removed.

Now, the system can be solvated in the next step. An optimally solvated box should look, as shown in Fig. 5.13.

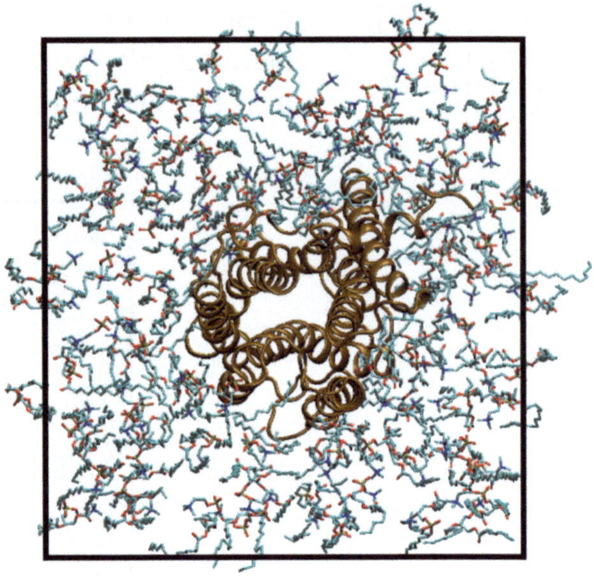

Fig. 5.12 Step 4: Equilibration of the lipid bilayer around the GPCR

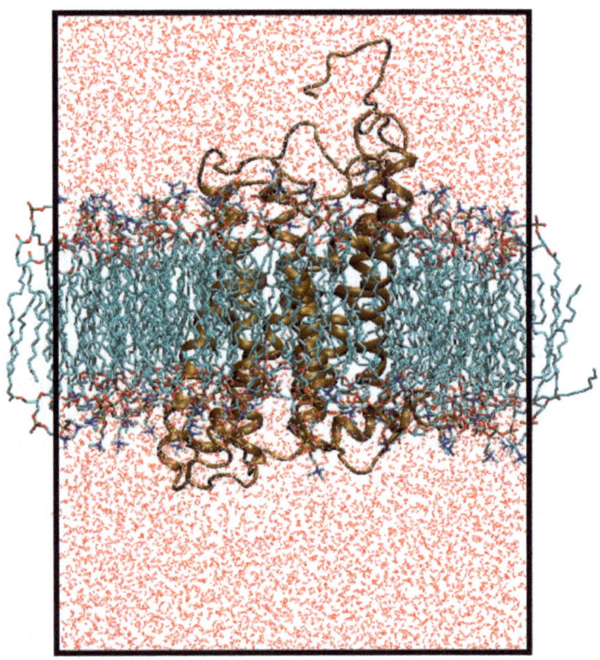

Fig. 5.13 Step 5: A well prepared simulation box, containing the GPCR, the lipid bilayer and extra- and intracellular water

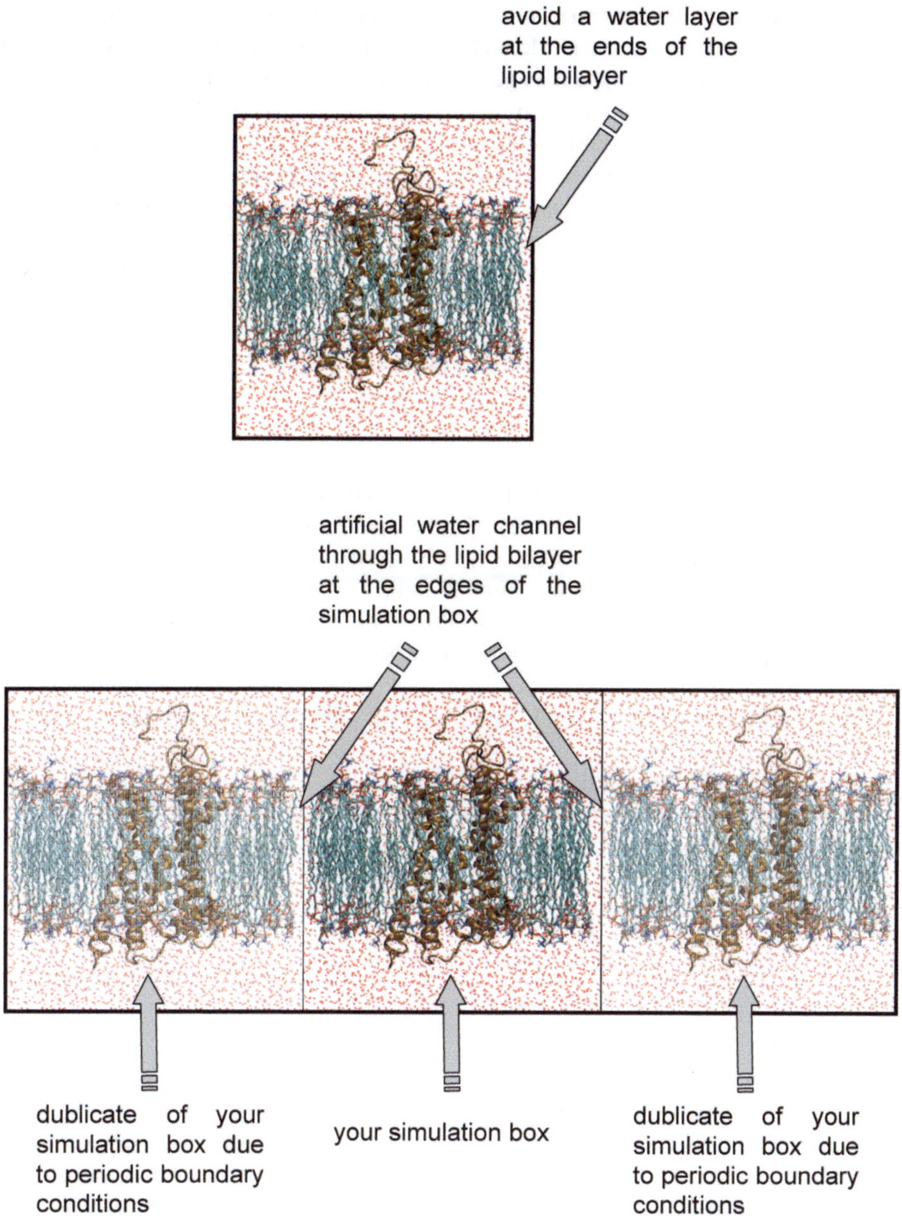

Fig. 5.14 Artificial water channels through the lipid bilayer at the edges of the simulation box as consequence of wrong system setup

With regard to solvation you have to look carefully onto the size of your simulation box: If the defined box size is a little bit larger, than the width of the lipid bilayer, and you perform the solvation, you may get another artefact, as shown in Fig. 5.14. Here, water channels/layers through the lipid bilayer are established. This is a completely wrong artefact and you should not do simulations with such systems. If you detect such a water channel/layer after solvation, you may remove the solvent, decrease the box size in an appropriate manner and solvate again. These steps should be repeated until the water channel/layer through the lipid bilayer is no longer observed.

After solvation, the system should be minimized using the GROMACS command `grompp` and `mdrun` (see Chap. 6). In the last step, the system has to be neutralized (see Chap. 6).

In the following box a short, stepwise summary of the alignment of a GPCR in the lipid bilayer is given.

- ☐ Construct a lipid bilayer or obtain it via download of a server
- ☐ Align your GPCR correctly into the lipid bilayer
- ☐ Remove the lipid molecules which overlap with the GPCR
- ☐ Center the system in the simulation box
- ☐ Minimize the system with GROMACS
- ☐ Equilibrate the lipids around the GPCR, position restraints should be put onto all sites of the protein using appropriate GROMACS commands
- ☐ Solvate your lipid-GPCR-complex with water in an appropriate manner (see also Chap. 6)
- ☐ Minimize the simulation box
- ☐ Neutralize the system and minimize again (see also Chap. 6)

Chapter 6
Minimization and Molecular Dynamics

A receptor model, which was energetically minimized, represents only one local minimum on the potential energy surface. Additionally, those minimized receptor models are based on homology models with more than 50 % difference in amino acid sequence compared to the template in most cases. Thus, receptor models should be refined by molecular dynamics (MD). Besides that, GPCRs, embedded in their natural surrounding, are not rigid, in contrast, they show a distinct flexibility. Thus, it is state of the art to analyze proteins by MD simulations (Carloni et al. 2002; Christen et al. 2008). In the early beginning of performing MD simulations of GPCRs the calculations were performed in gas phase without including the natural surrounding of the receptor. To avoid the destroy of the secondary and tertiary structure of the GPCR, position restraints were set onto the backbone of the transmembrane domains. However, this lead to wrong conformations of the amino acid sidechains, located at the surface of the receptor. To avoid such artefacts, the surrounding of the GPCR has to be included into the calculations. On the one hand, the surrounding stabilizes the conformation of the receptor. On the other hand, the correct surrounding allows the amino acid side chains on the receptor surface to achieve a correct conformation.

For enabling an adequate simulation box with the GPCR in its natural surrounding, at least four main steps, illustrated also in Fig. 6.1, have to be performed:

- ☐ Generate a complete model of the interesting GPCR
- ☐ Minimize the GPCR, position restraints should be put onto at least the backbone of the GPCR
- ☐ Put your GPCR correct into the lipid bilayer (see Chap. 5)
- ☐ Equilibrate the lipid bilayer around the GPCR, position restraints should be put onto at least the backbone of the GPCR
- ☐ If not already performed: center your system in the simulation box
- ☐ Solvate your lipid-GPCR-complex with water (see Chap. 5)
- ☐ Minimize your complete system; position restraints should be put onto at least the backbone of the GPCR
- ☐ Neutralize your simulation box to charge zero by putting an appropriate number of ions into the extra- or intracellular water

A. Strasser, H.-J. Wittmann, *Modelling of GPCRs*,
DOI 10.1007/978-94-007-4596-4_6, © Springer Science+Business Media Dordrecht 2013

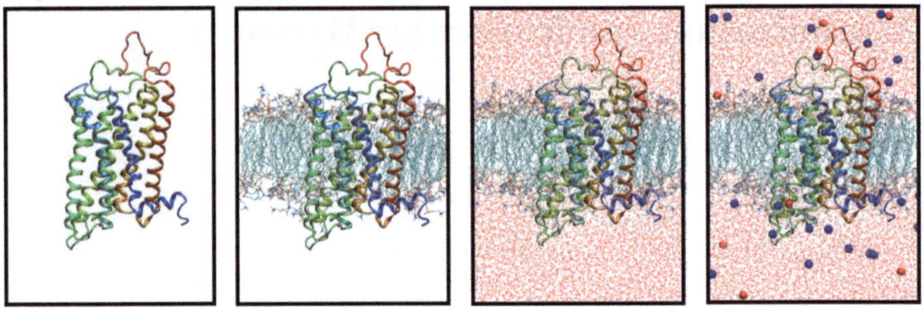

Fig. 6.1 Main steps for construction of a simulation box of a GPCR in the lipid bilayer

6.1 Generating a Complete Model of the Interesting GPCR

As already described in detail in Chap. 3, you should have designed a (homology) model of your GPCR and minimize the model in the gas phase (Fig. 6.1, step 1). In order to avoid destroying of the helical structure of the transmembrane domains, we recommend to set position constraints at least onto the backbone atoms. Therefore, you may use appropriate command of the GROMACS (http://www.gromacs.org) software package. But there is also a more flexible alternative in using LINUX-commands, as shown later on in Sect. 6.4.

6.2 Embedding the GPCR in a Lipid Bilayer

The embedding of the GPCR into a lipid bilayer (Fig. 6.1, step 2) is an important step, which has carried out very carefully. For a more detailed information see also Chap. 5.

6.3 Solvation of the Lipid-GPCR-Complex, Achiving Electroneutrality of the Simulation Box and Minimization

In the next step, the lipid-GPCR-complex should be solvated (Fig. 6.1, step 3). Some hints and pitfalls with regard to solvation of the lipid-GPCR-complex are mentioned in Chap. 5. Most modelling software allows an automatic solvation of your system. The solvation is very easy within GROMACS (http://www.gromacs.org). Here you can use the command `genbox`. If you have constructed a lipid-GPCR-complex in the file `rec_lipid.gro`, with the corresponding topology file `system.top`, you may perform the `genbox`-command for example like this:

```
> genbox -cp rec_lipid -cs -o rec_lipid_sol -p system↵
```

The option -cp is used to define the file, containing the structure, that should be solvated. The option -cs has to be used to define the solvent. With the option -o you define the name of your output file. Furthermore, we recommend to use the option -p and give the name of the topology file, you are already using. After completion of the genbox-command you should visualize your solvated system (here: rec_lipid_sol.gro) with an appropriate software, like vmd (http://www.ks.uiuc.edu/Research/vmd/). If your system looks like the example (Fig. 6.1, step 3), all is ok and you can go on with neutralizing your system. If your ligand or protein is outward of the water shell, you have to center the actual system in the simulation box using the editconf-command before performing the solvation process using the file rec_lipid.gro, containing the lipid-GPCR-complex:

```
> editconf -f rec_lipid.gro -c -o out.gro ↵
```

Rename the file out.gro to rec_lipid.gro with the help of the mv-command

```
> mv out.gro rec_lipid.gro ↵
```

Now, you may again perform the genbox-command, as mentioned above. If the resulting simulation box looks like the one in Fig. 5.13 everything worked well, but if it looks like Fig. 5.14, the reader is referred to Sect. 5.6.

After solvation, it is recommended, to minimize the system using the commands grompp and mdrun.

```
> grompp -f mini -c rec_lipid_sol -p system ↵
```

```
> mdrun -v -s ↵
```

An example parameter file mini.mdp, read by grompp is presented below.

```
;
;        mini.mdp
;
cpp                      = /lib/cpp
;define                  = -DPOSRES
constraints              = none
integrator               = steep
nsteps                   = 1000
;
;        Energy minimizing
;
emtol                    = 1000
emstep                   = 0.01
;
pbc                      = xyz
;
nstcomm                  = 1
```

```
nstlist            = 5
rlist              = 1.4
nstype             = grid
coulombtype        = pme
rcoulomb           = 1.4
epsilon_r          = 1.0
vdwtype            = Cut-off
rvdw               = 1.4
;DispCo            = EnerPres
Tcoupl             = no
Pcoupl             = no
gen_vel            = no
; Energy monitoring
energygrps         = system
```

Afterwards, you can start to neutralize your system (Fig. 6.1, step 4). To get information about the total charge of the system, have a look onto the output of the grompp command. Subsequently, you have to think about, which ions and how much you want to put into system. In general, sodium and chlorine ions are used. The concentration of sodium and chlorine ions should be chosen, that approximately physiological conditions are achieved.

Now you can neutralize your system using the command genion, as described in the GROMACS manual (van der Spoel et al. 2005).

After neutralization the system should be minimized again.

```
> grompp -f mini -c system -p system ↵
```

```
> mdrun -v -s ↵
```

If your system is minimized carefully and there are no "bugs", as described in Sect. 3.4.5, the MD simulation should work quite well.

6.4 Molecular Dynamic Simulation of your System

Now, the molecular dynamic simulation (van Gunsteren et al. 1990) can be started. In general, a MD simulation is divided into two phases: The equilibration phase and the productive phase. What does equilibration phase mean? Even if you put your GPCR very carefully in the lipid bilayer, the interactions between the lipid bilayer and the receptor are not very optimal, lets say, not equilibrated. Furthermore, during the solvation process, the water molecules are put somehow, of course in the correct density, around the lipid-GPCR-complex. But again, the interactions between the water molecules on the one hand and more importantly between the water molecules and the GPCR are not established. This means for example, no hydrogen bonds are established. If you start a molecular dynamic simulation without equilibration, the GPCR may be "destroyed", i.e. for example the helical conformation of the GPCR is not stable. In this case, your simulation results are wrong. In the equilibration phase, the surrounding of the GPCR, the lipid bilayer and the water, should be "equilibrated" around the GPCR without modifying the structure of the GPCR. This can be done, by putting position constraints onto the GPCR. Position restraints were already introduced in context with the minimization of the system. But it has to be taken into account, that in context with molecular dynamics, distinct "equilibration" protocols should be used, in order to perform a successful and well equilibration. At the beginning of the equilibration phase, a rather high force constant k_i is to be used, but during equilibration, the force constant should be decreased gradually, until a force constant of 0 is attained (Fig. 6.2).

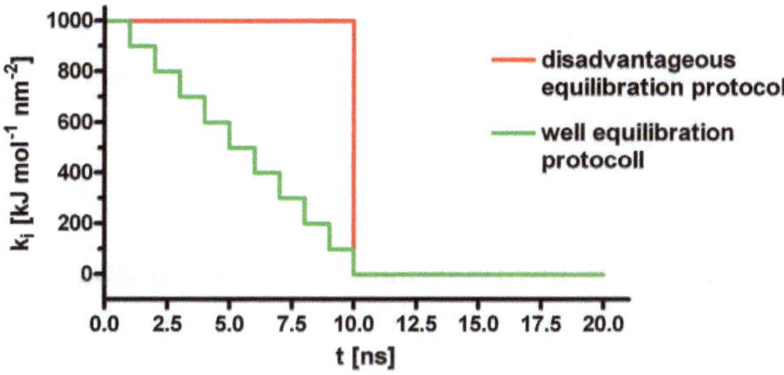

Fig. 6.2 Two different equilibration protocols for MD simulation

Of course, you can subsequently start each cycle of the equilibration protocol manually. However, it is more comfortable to establish a script `equilibrate_system`, which will be presented later on in this chapter.

First, one needs an appropriate position restraint file, which has the file-name-extension `itp` in general. Therefore one has to decide, which sites should be administered with position restraints. In the following you see a part of `gro`-file containing the coordinates of a protein in the ffG53a6-force-field notation. In the following example, the sites "C", "O", "N" and "H" should be administered with position restraints.

```
105SER      N   987    4.491    3.927    4.520
105SER      H   988    4.569    3.864    4.530
105SER     CA   989    4.495    4.049    4.604
105SER     CB   990    4.616    4.061    4.697
105SER     OG   991    4.739    4.063    4.624
105SER     HG   992    4.800    3.994    4.664
105SER      C   993    4.477    4.182    4.529
105SER      O   994    4.381    4.254    4.558
106MET      N   995    4.551    4.202    4.420
106MET      H   996    4.634    4.147    4.405
106MET     CA   997    4.533    4.321    4.333
106MET     CB   998    4.652    4.341    4.238
106MET     CG   999    4.788    4.340    4.309
106MET     SD  1000    4.803    4.453    4.452
106MET     CE  1001    4.950    4.385    4.525
```

```
106MET       C 1002   4.399   4.329   4.255
106MET       O 1003   4.341   4.437   4.244
107ASP       N 1004   4.344   4.213   4.214
107ASP       H 1005   4.396   4.128   4.216
107ASP      CA 1006   4.210   4.204   4.149
107ASP      CB 1007   4.189   4.061   4.099
107ASP      CG 1008   4.087   4.049   3.985
107ASP     OD1 1009   3.966   4.060   4.014
107ASP     OD2 1010   4.129   4.009   3.874
107ASP       C 1011   4.095   4.247   4.244
107ASP       O 1012   4.024   4.344   4.216
```

A GROMACS position restraint file starts with the keyword `[position_restraints]` followed by several lines. Each line corresponds to one site and contains five columns:

First column: Number of the site (numbering according to the topology file)
Second column: function type
Third column: force constant on the x-coordinate (kJ mol^{-1} nm^{-2})
Fourth column: force constant on the y-coordinate (kJ mol^{-1} nm^{-2})
Fifth column: force constant on the z-coordinate (kJ mol^{-1} nm^{-2})

Thus, at first, the number of the sites, which should be administered with position constraints has to be determined. The `gro`-file, which should be analyzed, is named `protein.gro`, for example. The numbers of the sites, administering with position restraints, should be written into the file `site.dat`:

```
> grep " C " protein.gro | cut -c 16-21 > site.dat↵
> grep " O " protein.gro | cut -c 16-21 >> site.dat↵
> grep " N " protein.gro | cut -c 16-21 >> site.dat↵
> grep " H " protein.gro | cut -c 16-21 >> site.dat↵
```

What does this sequence do? The command `grep " C " protein.gro` for example, looks for all lines in the file `protein.gro` which contain the string `" C "`, like shown below.

```
105SER    C    993   4.477   4.182   4.529
106MET    C   1002   4.399   4.329   4.255
107ASP    C   1011   4.095   4.247   4.244
```

Note, that only lines with a blank before and after the C are printed, because, the pattern for search is `" C "`. However, you do not see this output on your screen, because the results are connected via the pipe | to the command `cut`. Why is the command `cut` used? One needs not the complete line, but only the number of the site. If you have a closer look into `protein.gro`, you see, that the site numbers are written in the columns 17–20, if the protein contains not more than 9999 sites. The option "`-c 16-21`" cuts the columns 16–21 (including a blank before and after the site number) and redirects the results in to file `site.dat`. If you would use only one >, the file `site.dat` is created and the data are written into the new file. But be aware, if a file `site.dat` is already here in the current working directory, its data will be deleted. If the operator >> is used, all new data are appended to `site.dat`. Now, `site.dat` should contain the following information:

```
 993
1002
1011
```

After repeating the analogue commands with regard to O, N and H, the file site.dat should contain the following data:

```
 993
1002
1011
 994
1003
1012
 987
 995
1004
 998
 996
1005
```

Because the numbers are not sorted numerically, use the following command to ensure a correct order:

```
> sort -n site.dat > site_sort.dat ↵
```

To every site, a function type (second column) and a force constant for each coordinate (third to fifth column) has to be added. Therefore, we have to know, how much sites should be administered with position constraints. Because `site_sort.dat` does not contain any empty lines the appropriate number can be easily obtained using the command `wc`:

```
> wc -l site_sort.dat ↵
```

In actual example, there should be 12 lines. Thus, one has to create a new file containing "`1 1000 1000 1000`", if each force constant should have the value 1000, 12 times. This can be done using the following command:

```
> rm force.dat ↵
> set i = 1 ↵
> while ($i < = 12) ↵
>   echo "1 1000 1000 1000" >> force.dat ↵
>   @ i + + ↵
> end ↵
```

Now, both files, `site_sort.dat` and `force.dat` can be easily combined, using the command `paste`:

```
> echo "[position_restraints]" > posre_bb_1000.itp ↵
> paste site_sort.dat force.dat >> posre_bb_1000.itp ↵
```

If you performed all commands correctly, you should have the file `posre_bb_1000.itp` with the following data:

```
[position_restraints]
 987 1 1000 1000 1000
 993 1 1000 1000 1000
 994 1 1000 1000 1000
 995 1 1000 1000 1000
 996 1 1000 1000 1000
 998 1 1000 1000 1000
1002 1 1000 1000 1000
1003 1 1000 1000 1000
1004 1 1000 1000 1000
1005 1 1000 1000 1000
1011 1 1000 1000 1000
1012 1 1000 1000 1000
```

You see, that the command sequence, presented above, is very simple, in order to construct an appropriate file, containing information about position restraints. However, for your equilibration protocol, mentioned above, you will need several `itp`-files with different force constants. Therefore, the command sequence to generate the `itp`-file has to be repeated several times. Thus, it would be easier, to write an appropriate shell script.

```
#!/bin/tcsh

set fconst = (1000 800 600 400 200 100)
set nr_of_fconst = $#fconst

set i = 1

rm site.dat
rm force.dat

while ($i <= $nr_of_fconst)

grep " C " protein.gro | cut -c 16-21 >> site.dat
grep " O " protein.gro | cut -c 16-21 >> site.dat
grep " N " protein.gro | cut -c 16-21 >> site.dat
grep " H " protein.gro | cut -c 16-21 >> site.dat

sort -n site.dat > site_sort.dat
```

```

set nr_of_res = `wc -l site_sort.dat|
cut -d ' ' -f1`

set j = 1

while ($j <= $nr_of_res)
 echo "1 $fconst[$i] $fconst[$i] $fconst[$i] " >>
 force.dat
 @ j++
end

echo "[position_restraints]"> posre_bb_$fconst[$i].itp
paste site_sort.dat force.dat >>
posre_bb_$fconst[$i].itp

rm site.dat
rm force.dat

@ i++

end
```

You may name this shell script `gen_posre`. After saving the file ensure the execute permission by using the command:

```
> chmod u+x gen_posre ↵
```

Start your shell script, by typing

```
> gen_posre ↵
```

The contents of the new `itp`-files should be proofed using an editor. With this extensive example, you should see that the linux-commands, presented in the corresponding Chap. 11 are very useful in generating and handling large files. However, the lines above only represent a rudimentary shell script which can be expanded in order to be more flexible, like checking, if a file which has to be created, is already there in the directory. Actually, the script `gen_posre` does not take care about this. However, you can use and adopt the presented shell script `gen_posre` for your own purposes.

Take into account, that the first column in the `itp`-file has to contain the site numbers of the atoms, which have to be administered with position restraints. The numbering must be according to the numbering in the topology file! You can use the `gro`-file, as we did in our example, if you have only one protein and if the protein is the first "molecule" in your `gro`-file. If this is not the case, you are suggested to adopt the script `gen_posre` with regard to the topology file. Next distinct parts of a typical GROMACS topology-file, named `protein3.top` of a protein are shown:

```
[ moleculetype ]
; Name            nrexcl
Protein_3            3

[ atoms ]
;   nr       type  resnr residue  atom   cgnr     charge       mass  typeB  chargeB  massB
     1         NL      1    ALA      N      1      0.129    14.0067   ; qtot 0.129
     2          H      1    ALA     H1      1      0.248      1.008   ; qtot 0.377
     3          H      1    ALA     H2      1      0.248      1.008   ; qtot 0.625
     4          H      1    ALA     H3      1      0.248      1.008   ; qtot 0.873
     5        CH1      1    ALA     CA      2      0.127     13.019   ; qtot 1
     6        CH3      1    ALA     CB      2          0     15.035   ; qtot 1
     7          C      1    ALA      C      3       0.45     12.011   ; qtot 1.45
     8          O      1    ALA      O      3      -0.45    15.9994   ; qtot 1
     9          N      2    PRO      N      4          0    14.0067   ; qtot 1
    10        CH1      2    PRO     CA      5          0     13.019   ; qtot 1
    11       CH2R      2    PRO     CB      5          0     14.027   ; qtot 1
    12       CH2R      2    PRO     CG      6          0     14.027   ; qtot 1
    13       CH2R      2    PRO     CD      6          0     14.027   ; qtot 1
    14          C      2    PRO      C      7       0.45     12.011   ; qtot 1.45
    15          O      2    PRO      O      7      -0.45    15.9994   ; qtot 1
    16          N      3    GLY      N      8      -0.31    14.0067   ; qtot 0.69
    17          H      3    GLY      H      8       0.31      1.008   ; qtot 1
    18        CH2      3    GLY     CA      9          0     14.027   ; qtot 1
    19          C      3    GLY      C     10       0.45     12.011   ; qtot 1.45
    20          O      3    GLY      O     10      -0.45    15.9994   ; qtot 1
    21          N      4   CYSH      N     11      -0.31    14.0067   ; qtot 0.69
    22          H      4   CYSH      H     11       0.31      1.008   ; qtot 1
    23        CH1      4   CYSH     CA     12          0     13.019   ; qtot 1
    24        CH2      4   CYSH     CB     13       0.15     14.027   ; qtot 1.15
    25          S      4   CYSH     SG     13      -0.37      32.06   ; qtot 0.78
    26          H      4   CYSH     HG     13       0.22      1.008   ; qtot 1
    27          C      4   CYSH      C     14       0.45     12.011   ; qtot 1.45
    28          O      4   CYSH      O     14      -0.45    15.9994   ; qtot 1
    29          N      5    GLY      N     15      -0.31    14.0067   ; qtot 0.69
    30          H      5    GLY      H     15       0.31      1.008   ; qtot 1
    31        CH2      5    GLY     CA     16          0     14.027   ; qtot 1
    32          C      5    GLY      C     17       0.45     12.011   ; qtot 1.45
    33          O      5    GLY      O     17      -0.45    15.9994   ; qtot 1
    34          N      6    ALA      N     18      -0.31    14.0067   ; qtot 0.69
    35          H      6    ALA      H     18       0.31      1.008   ; qtot 1
    36        CH1      6    ALA     CA     19          0     13.019   ; qtot 1
...
   439          C     52    LEU      C    193       0.45     12.011   ; qtot 6.45
   440          O     52    LEU      O    193      -0.45    15.9994   ; qtot 6
   441          N     53   HISB      N    194      -0.31    14.0067   ; qtot 5.69
   442          H     53   HISB      H    194       0.31      1.008   ; qtot 6
   443        CH1     53   HISB     CA    195          0     13.019   ; qtot 6
   444        CH2     53   HISB     CB    195          0     14.027   ; qtot 6
   445          C     53   HISB     CG    196          0     12.011   ; qtot 6
   446         NR     53   HISB    ND1    196      -0.54    14.0067   ; qtot 5.46
   447        CR1     53   HISB    CD2    196       0.14     13.019   ; qtot 5.6
   448        CR1     53   HISB    CE1    196       0.14     13.019   ; qtot 5.74
   449         NR     53   HISB    NE2    196      -0.05    14.0067   ; qtot 5.69
   450          H     53   HISB    HE2    196       0.31      1.008   ; qtot 6
   451          C     53   HISB      C    197       0.45     12.011   ; qtot 6.45
   452          O     53   HISB      O    197      -0.45    15.9994   ; qtot 6
   453          N     54    VAL      N    198      -0.31    14.0067   ; qtot 5.69
   454          H     54    VAL      H    198       0.31      1.008   ; qtot 6
   455        CH1     54    VAL     CA    199          0     13.019   ; qtot 6
   456        CH1     54    VAL     CB    199          0     13.019   ; qtot 6
   457        CH3     54    VAL    CG1    199          0     15.035   ; qtot 6
   458        CH3     54    VAL    CG2    199          0     15.035   ; qtot 6
   459          C     54    VAL      C    200       0.27     12.011   ; qtot 6.27
   460         OM     54    VAL     O1    200     -0.635    15.9994   ; qtot 5.635
   461         OM     54    VAL     O2    200     -0.635    15.9994   ; qtot 5

[ bonds ]
;  ai    aj funct            c0            c1            c2            c3
    1     2     2    gb_2
    1     3     2    gb_2
...
```

This topology file also consists of all information, which is needed for construction of a position restraint-file. The protein consists of 461 sites, which are defined from line 7–467. Thus, to extract information with regard to site number and atom, the lines 7–467 are important and they can be obtained via the command line:

```
> head -n 467 protein3.top | tail -n 461↵
```

If you perform the command, as shown above, you get the output containing 461 lines onto your xterm. However, we are not interested for the whole information of a line. Instead, if only backbone atoms should be administered with position restraints, we have to look for the corresponding site numbers (column title: `nr`) of the backbone atoms (column title: `atom`), using the following sequence of commands:

```
> head -n 467 protein3.top | tail -n 461 | tr -s ' ' |
  cut -d ' ' -f2,6 | grep ' C$' | cut -d' ' -f1 >
  site.dat↵

> head -n 467 protein3.top | tail -n 461 | tr -s ' ' |
  cut -d' ' -f2,6 | grep ' O$' | cut -d' ' -f1 >>
  site.dat↵

> head -n 467 protein3.top | tail -n 461 | tr -s ' ' |
  cut -d ' '-f2,6 | grep ' N$ ' | cut -d' ' -f1 >>
  site.dat↵

> head -n 467 protein3.top | tail -n 461 | tr -s ' ' |
  cut -d ' '-f2,6 | grep ' H$' | cut -d' ' -f1 >>
  site.dat↵
```

The output of the `head`- and `tail`-command is directed via pipe to the command `tr`. The command `tr` with the option `-s ' '` combines all subsequent white space characters to exactly one. For example

```
echo "xxx        xxx" | tr -s ' '
```
outputs: `xxx xxx`

Thus, line 7, containing information about site 1, may look like that, after using the command `tr -s` as described above:

```
 1 NL 1 ALA N 1 0.129 14.0067; qtot 0.129
```

Due to the white space character in column 1, column 2 and 6 are of interest for us: Column 2 in the line above contains information about the site and column 6 in the line above contains information about the type. Thus, the command

```
> head -n 467 protein3.top | tail -n 461 | tr -s ' ' |
cut -d' ' -f2,6↵
```

would lead to the following output (only the first seven lines are shown):

```
1 N
2 CA
3 CB
4 C
5 O
6 N
7 CA
```

Now, we have to look for all lines containing the sites, which should be administered with position restraints. In our case, this is C, O, N and H. This can be achieved by combining the command, explained above, with a corresponding `grep` command, as shown below:

```
>head -n 467 protein3.top | tail -n 461 | tr -s ' ' |
cut -d ' '-f2,6 | grep ' C$'↵
```

This command leads to the following output (only the first ten lines are shown):

```
7 C
```

Please compare the option of `grep` with the options, which were used, when dealing the same problem with the `gro`-file. In the `gro`-file, the search string could be defined as `" C "`. This means, that grep searched all lines, containing a `C` with a blank before and after the `C`. But in the actual case, one has to be aware, that there is a blank before the `C`, but there is no blank after the `C`, because, the line ends with a new line. Thus, if one searches for `"C"`, all lines with `"C"`, but also with `"CA"` and `"CB"` for example, were found. In order to avoid this, a new search criterion has to be found. This might be: Look for all lines containing a `C` at the end of a line and with a blank before the `C`. The can be achieved by `grep ' C$'`, as shown above. The `$` after the search string induces, that `grep` only searches the string at the end of a line. In order to avoid that the $ is misinterpreted as variable substitution, the single quotes have to be used instead of double quotes.

For the position restraints, only the number of the corresponding sites is of interest, thus, the long command line above has to be combined at last with the `cut`-command in the following manner:

```
> head -n 467 protein3.top | tail -n 461 | tr -s ' ' |
  cut -d' ' -f2,6 | grep ' C$' | cut -d' ' -f1↵
```

The further steps in handling the file `site.dat` are the same, as already mentioned above.

Supposing the existence of the constraint files created above, the following shell-script `equilibrate_system` can be used for equilibration of the simulation box. Be aware that the files `system.top`, `system.gro` (minimized simulation box, see Sect. 6.3), `md_first.mdp` (mdp-file for the first equilibration cycle), `md.mdp` (mdp-file for all following cycles) and the `itp`-files reside in the same directory as the shell-script.

```
#!/bin/tcsh -f

set fconst = (1000 800 600 400 200 100)

set nr_of_fconst = $#fconst

set i = 1

while ($i <= $nr_of_fconst)
  mkdir posre_${i}
  cd posre_${i}
  cp ../system.top .
  cp ../posre_bb_$fconst[$i].itp ./posre.itp

  if ($i == 1) then
    cp ../system.gro .
    cp ../md_first.mdp .
    grompp -f md_first -o md_first -c system
     -p system
    wait
    mdrun -v -s md_first -e md_first -o md_first
     -c after_md -g shortlog
    wait
  else
    cp ../md.mdp .
    @ k = $i - 1
    cp ../$posre_$k/after_md.gro ./system.gro
    grompp -f md -o md -c system -p system
    wait
    mdrun -v -s md -e md -o md -c after_md
     -g shortlog
    wait
  endif
  cd ..
  @ i++
end
```

The `grompp` input file `md_first.mdp` with exemplary parameters is shown below:

```
;       md_first.mdp
;       MD
;
;       Input file
;
title                       = System
cpp                         = /lib/cpp
define                      = -DPOSRES
;constraints                = all-bonds
;constraint_algorithm       = lincs
unconstrained_start         = yes
integrator                  = md
tinit                       = 0
dt                          = 0.001; ps!
nsteps                      = 100000
nstcomm                     = 1
; Output control
nstxout                     = 5000
nstvout                     = 5000
nstfout                     = 0
nstlog                      = 5000
nstenergy                   = 100
; Neighbor searching
nstlist                     = 10
ns_type                     = grid
pbc                         = xyz
rlist                       = 1.4
; Electrostatics and VdW
coulombtype                 = PME
;rcoulomb_switch            = 0
rcoulomb                    = 1.4
epsilon_r                   = 1.0
;epsilon_rf                 = 7.0
vdwtype                     = Cut-off
;rvdw_switch                = 0
rvdw                        = 1.4
;DispCorr                   = EnerPres
fourierspacing              = 0.12
fourier_nx                  = 0
fourier_ny                  = 0
fourier_nz                  = 0
pme_order                   = 4
ewald_rtol                  = 1e-5
optimize_fft                = yes
```

```
; Temperature coupling
tcoupl                  = berendsen
tc-grps                 = system
tau_t                   = 0.1
ref_t                   = 298
; Energy monitoring
energygrps              = system
; Pressure coupling is not on
Pcoupl                  = berendsen
pcoupltype              = isotropic
tau_p           = 0.5 0.5 0.5 0.0 0.0 0.0
compressibility = 4.5e-5 4.5e-5 4.5e-5 0.0 0.0 0.0
ref_p           = 1.0
; Generate velocites is on at 298 K.
gen_vel                 = yes
gen_temp                = 298
gen_seed                = 173529
```

In the file `md.mdp`, the parameters `unconstrained_start` and `gen_vel` should be set `no`. Afterwards, the productive simulation phase without position restraints can be started.

If the binding-mode of a ligand-receptor-complex should be analyzed via MD simulations, analogous steps, as shown above, have to be performed. Often, it is very useful, to administer the ligand with an equilibration protocol, similar to that, describe above for the receptor.

For analysis of the MD simulation, several GROMACS commands, like `g_energy`, `g_hbond`, `g_rms` and `g_traj`, for example, can be used.

It has to be taken into account, that water molecules can penetrate into the binding-pocket and mediate interactions between the ligand and receptor, as illustrated in Fig. 6.3.

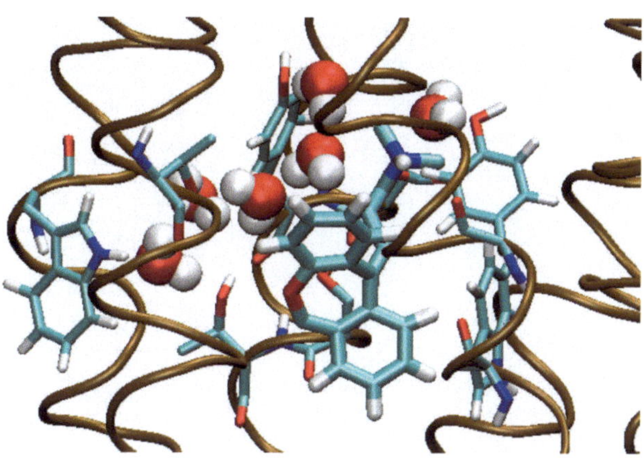

Fig. 6.3 Internal water molecules mediate the interaction between ligand and receptor

Chapter 7
Calculation of Gibbs Energy of Solvation

7.1 Theory – Link Between Microscopic and Macroscopic World

In the next chapters, a short summary with regard to link the microscopic and macroscopic world is given. For a more detailed description, the reader is referred to the literature (van Gunsteren and Berendsen 1987; Jensen 1999; Frenkel and Smit 2002; van der Spoel et al. 2005).

7.1.1 *Statistical Mechanical Basics*

In this chapter we deal with the problem to connect a model or a microscopic picture of matter to measurable macroscopic quantities. Linking these two worlds represents the only possibility to validate models and gain insight into the molecular processes. Referring to the chapter of thermodynamical basics we established a model for the ligand receptor interaction by formulating the equilibrium

$$\mathrm{L} + \mathrm{R} \rightleftharpoons \mathrm{LR} \tag{7.1}$$

characterized by its equilibrium constant, a measurable quantity:

$$K = \frac{c_{LR} c_o}{c_L c_R} \tag{7.2}$$

To understand the processes leading to this equilibrium constant on a molecular level, we remember the fundamental equation resulting from the first and second law of thermodynamics in the case of constant pressure and temperature:

$$\Delta G^o = -RT \ln K. \tag{7.3}$$

Because we transfer the measurable quantity K into the energetic quantity ΔG^o we make the first move to answer our central question. Obviously the next step is the

A. Strasser, H.-J. Wittmann, *Modelling of GPCRs*,
DOI 10.1007/978-94-007-4596-4_7, © Springer Science+Business Media Dordrecht 2013

connection of ΔG^o to the interactions which take place when the ligand L leaves its solvation state and enters the receptor R to form the complex LR. This link is given by the concepts of Classical Statistical Mechanics in combination with Quantum Mechanics. As formerly stated (Chap. 1), Quantum Mechanics would be the best choice for describing the behaviour of matter in a microscopic world but up to now it is impossible to handle large biochemical systems. So we use the Classical Statistical Mechanics which uses the Hamiltonian function

$$H(\vec{p},\vec{r}) = E_{kin}(\vec{p}) + E_{pot}(\vec{r}) \tag{7.4}$$

the sum of the total kinetic (E_{kin}) and the potential energy (E_{pot}) as a central function to calculate macroscopic quantities. H depends on the momenta ($\vec{p}$) and the coordinates ($\vec{r}$) of all species present in the system of interest. Because we are interested in the equilibrium state of a system, H depends not explicitly on time. The expression for the kinetic part of H is the sum over the kinetic energies of all species i:

$$E_{kin}(\vec{p}) = \sum_i \frac{\vec{p}_i^{\,2}}{2m_i} \tag{7.5}$$

where m_i denotes the mass and $\vec{p}_i$ the momentum of a particle i. The potential energy $E_{pot}(\vec{r})$ comprises energies resulting from binding interactions, ion-ion, ion-dipole or dipole-dipole interactions. Note, that the Hamiltonian function does not contain variables like the pressure p or the temperature T. These parameters appear in the expressions of macroscopic quantities like the internal energy U of a system with fixed volume, temperature and number of particles, which will read as:

$$U(V,T,N) = \frac{\sigma_V}{Q_V} \int \ldots \int H\left(\vec{p}_i,\vec{r}_i\right) \exp\left(-\frac{H\left(\vec{p}_i,\vec{r}_i\right)}{kT}\right) d\vec{p}_i d\vec{r}_i \tag{7.6}$$

where

$$Q_V = \sigma_V \int \ldots \int \exp\left(-\frac{H\left(\vec{p}_i,\vec{r}_i\right)}{kT}\right) d\vec{p}_i d\vec{r}_i \tag{7.7}$$

k and T denote the Boltzmann constant and the temperature, N is the total number of particles in the system and σ_V is a normalisation constant. The integration is extended over all values of the momenta and coordinates of all species i present in the system. The quantity Q_V is called the partition function at constant volume. Having a closer look onto the Eqs. 7.6 and 7.7 U is identified as the mean value of the Hamiltonian function H over the so called phase space given by all the momenta and coordinates. Referring to a system with fixed pressure, temperature and number of particles we get:

$$\begin{aligned} H(p,T,N) = \frac{\sigma_p}{Q_p} \int \int \ldots \int \left(H\left(\vec{p}_i,\vec{r}_i\right) + pV\right) \\ \cdot \exp\left(-\frac{H\left(\vec{p}_i,\vec{r}_i\right) + pV}{kT}\right) d\vec{p}_i d\vec{r}_i dV \end{aligned} \tag{7.8}$$

where

$$Q_p = \sigma_p \int \int \dots \int \exp\left(-\frac{H(\vec{p}_i, \vec{r}_i) + pV}{kT}\right) d\vec{p}_i d\vec{r}_i dV \tag{7.9}$$

is the partition function at constant pressure with the normalisation constant σ_p. Equations 7.8 and 7.9 contain the product of the pressure p and the volume V to transform the internal energy U into the enthalpy H, which must not be mixed up with the Hamiltonian function $H(\vec{p}, \vec{r})$. The Gibbs energy for a system at constant pressure, temperature and number of particles reads:

$$G = -kT \ln Q_p. \tag{7.10}$$

It should be taken into consideration that the Gibbs energy G does not represent a mean value, like the internal energy U or the enthalpy H, which will lead to new concepts in the calculation of this quantity in the framework of Molecular Dynamics.

So, making use of the concepts of Statistical Mechanics we are able to calculate all important thermodynamical quantities after formulating the potential energy E_{pot}. We therefore should be able to calculate measurable quantities like the ligand receptor association constant and compare the results with experimental values in order to refine our models.

7.1.2 From Potential Energy to the Chemical Potential

To calculate the equilibrium constant for the association process, Eq. 7.1, we remember the equation from the chapter of thermodynamical basics:

$$\Delta G^o = \mu_{LR}^o - \mu_L^o - \mu_R^o. \tag{7.11}$$

Because the chemical potential of each species *LR*, *L* or *R* means its Gibbs energy per mole in accordance to Eq. 7.10 we have to evaluate the partition function Q_p for constant pressure and constant temperature of an appropriate system, containing a number of particles *LR*, *L* or *R*, which equals the Avogadro number N_A, considering the particular reference state. After defining the potential energy $E_{pot}(\vec{r})$ we have to integrate over the whole phase space, which indeed is a very difficult task as this function will not be given in a simple analytical form. There are attempts to solve this problem with the so-called Monte Carlo method (Metropolis 1987; Bouzida et al. 1992), but this procedure takes a lot of time and the results very often are not satisfactory. For large biochemical systems, the Molecular Dynamics is a widely accepted alternative (Christen and van Gunsteren 2008). This concept uses the Newton equation of motion to follow the evolution of an arbitrary system with time, i. e. we will monitor all properties of interest, for example the potential and kinetic energy of our system, in their dependence of time. But now another problem arises in defining the equilibrium state and calculating the corresponding macroscopic quantities U, H and G.

First we postulate the equality between the mean values in phase space and the mean on time scale from MD calculations, which is applicable for the quantities U and H. Having a look on Eq. 7.10 we see that this concept does not hold for the Gibbs energy. So there is no simple possibility to calculate this important quantity and therefore no way to estimate the equilibrium constant on the base of Molecular Dynamics. To overcome this problem, the concept of perturbation with respect to the potential energy is introduced. Suppose we have a system at constant pressure, temperature and constant number of particles. We indicate this state as state "1" and according to Eq. 7.9 the partition function $Q_p(1)$ for an arbitrary species is given by the equation:

$$Q_p(1) = \int\int\ldots\int \exp\left(-\frac{H(1)\,(\vec{p}_i,\vec{r}_i) + pV}{kT}\right) d\vec{p}_i d\vec{r}_i dV. \tag{7.12}$$

Now assume the Hamiltonian energy of our system has changed to state "2" by variation of the particle interactions for example. We will then write the partition function $Q_p(2)$:

$$Q_p(2) = \int\int\ldots\int \exp\left(-\frac{H(2)\,(\vec{p}_i,\vec{r}_i) + pV}{kT}\right) d\vec{p}_i d\vec{r}_i dV. \tag{7.13}$$

Calculating the Gibbs energy for example of state "2" according to Eq. 7.10 yields:

$$G(2) = -kT \ln Q_p(2). \tag{7.14}$$

Let us reformulate this equation in the following way:

$$G(2) = -kT \ln \left(\frac{Q_p(1) Q_p(2)}{Q_p(1)}\right). \tag{7.15}$$

To what extent does this mathematical manipulation help us to solve our problem? First we can write the right hand side of Eq. 7.15 in the following form:

$$G(2) = -kT \ln (Q_p(1)) - kT \ln \frac{Q_p(2)}{Q_p(1)} \tag{7.16}$$

and making use of Eq. 7.10, we have

$$G(2) = G(1) - kT \ln \frac{Q_p(2)}{Q_p(1)}. \tag{7.17}$$

Next we will reformulate $Q_p(2)$:

$$\begin{aligned} Q_p(2) = \int\int\ldots\int \exp\left(-\frac{H(2)\,(\vec{p}_i,\vec{r}_i) + pV - (H(1)\,(\vec{p}_i,\vec{r}_i) + pV)}{kT}\right) \\ \cdot \exp\left(-\frac{H(1)\,(\vec{p}_i,\vec{r}_i) + pV}{kT}\right) d\vec{p}_i d\vec{r}_i dV \end{aligned} \tag{7.18}$$

which only means to multiply the integrand of Eq. 7.13 by one. Now, the resultant term $Q_{p(2)}/Q_{p(1)}$ represents the phase mean of the quantity

$$\exp\left(-\frac{H(2)\,(\vec{p}_i,\vec{r}_i)+pV-(H(1)\,(\vec{p}_i,\vec{r}_i)+pV)}{kT}\right). \tag{7.19}$$

This result is very important for the Molecular Dynamics simulation procedure because this phase mean corresponds to the mean in time scale. We follow the time evolution of our system in state "1" and calculate $H(2)$ for the same set of variables $\vec{p}_i$ and $\vec{r}_i$. The result represents the difference $G(2){-}G(1)$ due to a change in the potential energy between the states 1 and 2. So we are able to use the MD simulation method to calculate a difference of the Gibbs energy for two distinct states of our system with the help of the so-called thermodynamic perturbation formula.

7.1.3 The Concept of the Coupling Parameter Within MD Simulations

The mentioned procedure for calculating the mean according to Eq. 7.18 sometimes leads to convergence problems in MD simulations especially if state "2" energetically is far from state "1". As a workaround to get the term $G(2){-}G(1)$ a stepwise integration based on small differences of the Hamiltonian energies between neighbouring states could be performed. Returning to Eq. 7.10 we introduce a so-called coupling parameter λ which described the state of the system. Therefore, a variation of λ indicates a change of the system state and we can write

$$G(\lambda)=-kT\ln\left(Q_p(\lambda)\right) \tag{7.20}$$

where $Q_p(\lambda)$ reads

$$Q_p(\lambda)=\int\int\cdots\int\exp\left(-\frac{H(\lambda)\,(\vec{p}_i,\vec{r}_i)+p\cdot V}{k\cdot T}\right)d\vec{p}_i d\vec{r}_i dV. \tag{7.21}$$

Within the framework of the coupling-parameter concept, the Coulomb potential for instance between two sites i and j separated by the distance r_{ij} for example, is defined by the following equation:

$$E_{el}=\frac{1}{4\pi\varepsilon_o r_{ij}}\left((1-\lambda)q_i^{(1)}q_j^{(1)}+\lambda q_i^{(2)}q_j^{(2)}\right). \tag{7.22}$$

Therein, the superscripts (1) and (2) refer to the state 1 and 2, respectively.

A differential small change in λ, $d\lambda$ will lead to the following expression for Q_p:

$$\frac{dQ_p(\lambda)}{d\lambda} = -\frac{1}{kT}\int\int\cdots\int \frac{dH(\lambda)(\vec{p}_i,\vec{r}_i)}{d\lambda} \cdot \exp\left(-\frac{H(\lambda)(\vec{p}_i,\vec{r}_i)+pV}{kT}\right) d\vec{p}_i d\vec{r}_i dV \tag{7.23}$$

and according to Eq. 7.10

$$\frac{dG(\lambda)}{d\lambda} = -\frac{kT}{Q_p(\lambda)}\frac{dQ_p(\lambda)}{d\lambda} \tag{7.24}$$

where the term $\frac{dQ(\lambda)}{d\lambda}\frac{1}{Q_p(\lambda)}$ represents a mean value of $^{dH(\lambda)(\vec{p}_i,\vec{r}_i)}/_{d\lambda}$ and so we are able to calculate $^{dG(\lambda)}/_{d\lambda}$ as a function of λ by MD simulation. A numerical integration procedure will yield the desired term $G(2)-G(1)$. The concept of the coupling parameter λ could be thought as special case of the energy perturbation concept mentioned in the foregoing section.

In the next two sections, we will apply the concept of the coupling parameter to the calculation of the Gibbs energy of solvation for the system ethanol/water and subsequently to the estimation of the equilibrium constant for the ligand binding process.

7.2 Examples – Conceptual and Practical Considerations

7.2.1 Example 1: Ethanol in Water – Conceptual Considerations

From a thermodynamic point of view, the desired solvation energy requires the calculation of the reference chemical potentials (at the pressure of 1 bar and the temperature of 298.15 K), when transferring one mole of ethanol from the ideal gas state into the solvent (water), forming an ideal solution of concentration 1 mol/l.

$$\Delta G^o_{sol} = \mu^o_{EtOH} - \mu^{o,g}_{EtOH}. \tag{7.25}$$

Applying the coupling parameter concept, we will start from a system state 1, which corresponds to the solution of n_{EtOH} moles of ethanol and n_{H_2O} moles of water. In the next step, we will switch off all interactions between ethanol and water, arriving in a system state 2, comprising of the pure solvent and a "solute", which will be considered as an ideal gas. The Gibbs energy corresponding to state 1 will read:

$$G(1) = n_{EtOH}\left(\mu^o_{EtOH} + RT\ln\frac{c_{EtOH}}{c_o}\right) + n_{H_2O}\left(\mu^*_{H_2O} + RT\ln x_{H_2O}\right). \tag{7.26}$$

Here, we assume a dilute solution of ethanol in water in order to neglect the interactions between the solute molecules and the influence of remaining ethanol molecules on the interaction of a particular solute molecule with the solvent. This assumption allows omitting the activity coefficients (compare Chap. 10) and so we are able to establish an ideal solution. The reference chemical potential of the solvent $\mu^*_{H_2O}$ refers to the pure solvent at 1 bar and 298.15 K. The corresponding concentration variable is then given by the mole fraction x_{H_2O}.

For state 2, we may write:

$$G(2) = n_{EtOH}\mu^{o,g}_{EtOH} + n_{H_2O}\mu^*_{H_2O} \tag{7.27}$$

where $\mu^{o,g}_{EtOH}$ denotes the reference state of the ideal gas ethanol at standard pressure. Taking the difference $G(1)-G(2)$ we arrive at:

$$G(1) - G(2) = n_{EtOH}\left(\mu^o_{EtOH} - \mu^{o,g}_{EtOH}\right) + n_{EtOH}RT\ln\frac{c_{EtOH}}{c_o} + n_{H_2O}RT\ln x_{H_2O} \tag{7.28}$$

because the terms containing the reference chemical potential for the solvent cancel and the first term within parenthesis of the right hand side of Eq. 7.28 corresponds to the Gibbs energy of solvation. The left hand side of Eq. 7.28 will be calculated with the help of MD-coupling-parameter concept and so requires to set up an appropriate system. In the framework of GROMACS MD simulations, the quantity mole may be replaced by particle numbers. Thus, one mole of some species means one particle of the species in the simulation. To set up the system, one would choose one molecule of ethanol, firstly, to fulfil the requirement of neglecting the formerly discussed interactions, and secondly, to have the desired term $\mu^o_{EtOH} - \mu^{o,g}_{EtOH}$ isolated on the right hand side of Eq. 7.28. The next problem is the choice of the quantity n_{H_2O}. Of course, one could try to use a number of water molecules in such a way, that the two concentration terms on the right hand side of Eq. 7.28 would cancel:

$$RT\left(\ln\frac{c_{EtOH}}{c_o} + n_{H_2O}\ln\frac{n_{H_2O}}{1\text{ mol} + n_{H_2O}}\right) = 0. \tag{7.29}$$

The calculated difference $G(1)-G(2)$ equals the Gibbs energy of solvation in this case. The concentration term c_{EtOH} may be written approximately as

$$c_{EtOH} = \frac{n_{EtOH}\rho_{H_2O}}{n_{EtOH}M_{EtOH} + n_{H_2O}M_{H_2O}} \tag{7.30}$$

where the symbols M_{EtOH} and M_{H_2O} denote the molar masses of the solute and the solvent, ρ_{H_2O} means the density of water and will be substituted in the case of dilute solution by 1 kg/l. Taking into account $n_{EtOH} = 1$ mol, we get an equation for estimating n_{H_2O}, which, after solving numerically, exhibits the result of 18 water molecules. However, for a simulation with periodic boundary conditions, the system comprised of 19 molecules will be too small. So, the question arises, how many water molecules to use and how to treat the concentration terms in Eq. 7.28. To set up a system large enough for an exact simulation, the box size should be as large as possible, i.e. n_{H_2O} is much larger than n_{EtOH}, x_{H_2O} will be approximately one and the term $\ln x_{H_2O}$ will reach zero. But if we choose $n_{EtOH} = 1$ mol and calculate the term $n_{H_2O}RT \ln x_{H_2O}$ for several numbers of water molecules, we would conjecture the limiting value $-RT$. So, what is the correct result? As we demand a large number of solvent particles, we already see that the logarithm of the mole fraction of the solvent reaches the value zero, but this term is multiplied by the n_{H_2O} itself. On the one hand, we have a term approaching zero, but on the other hand we have a term getting larger and larger. To achieve the correct result of this product, we have to use the so-called rule of L'Hospital which tells us that the limiting value of the expression will be $-RT$. If we choose $N_{H_2O} = 512$, we get $x_{H_2O} = 0.998051$ and $n_{H_2O}RT \ln x_{H_2O} = -2.476$ kJ per mol ethanol. For $N_{H_2O} = 3{,}000$, $x_{H_2O} = 0.999667$ and $n_{H_2O}RT \ln x_{H_2O} = -2.477$ kJ per mole ethanol, which is close to the limiting value $-RT$ (-2.479 kJ per mole ethanol) at a temperature of 298.15 K. So, a choice of the number of solvent molecules between 512 and 3,000 or higher will lead to a constant term of approximately $-RT$ in Eq. 7.28. Now let us have a look onto the concentration term in Eq. 7.28:

$$n_{EtOH}RT \ln \frac{c_{EtOH}}{c_o}. \tag{7.31}$$

Setting n_{EtOH} equal to 1 mol, the value of c_{EtOH} is determined by the choice of the number of water molecules, which define the system volume. Having $N_{H_2O} = 800$, we get a cubic box size of 2.87538 nm at $T = 298.15$ K and $p = 1$ bar. The concentration of ethanol for a box volume V is then given by

$$c_{EtOH} = \frac{1\ \text{mol}}{N_A V} \tag{7.32}$$

with a value of 0.0698 mol/l. The corresponding energy term for 1 mol of ethanol will become $RT \ln \frac{c_{EtOH}}{c_o} = -6.599$ kJ. The experimental value of ΔG^o_{sol} is -20.98 kJ/mol. Thus, the concentration term just calculated, represents a fraction of approximate 30 % and the corresponding solvent term of about 12 %. To reduce the amount of corrections necessary for calculating solvation energies, Villa et al has given a workaround of this problem by taking the number of solvent molecules as 3,000, placed in a cubic box of length 4.5 nm, using a so-called twin-range cut-off distance of 0.8 and 1.4 nm, respectively (Villa and Mark 2002).

The partial charges of the sites of the solutes are adjusted to reproduce the experimental values of the quantity ΔG^o_{sol} neglecting the discussed concentration terms. So the calculated difference $G(1)-G(2)$ directly corresponds to the Gibbs energy of solvation. Referring to our present example, a number of 3,000 water molecules and 1 mol molecule of ethanol gives rise to a mol fraction of water $x_{H_2O} = 0.999667$, whereas the molar concentration of ethanol reads as $c_{EtOH} = 0.01822$ mol/l. At a temperature of 298.15 K, the concentration terms of Eq. 7.28 will influence the above mentioned difference of the Gibbs energies by $RT\left(\ln \frac{c_{EtOH}}{c_o} + n_{H_2O} \ln x_{H_2O}\right) = -12.41$ kJ per mol ethanol, which is about 60 % of the experimental value

Table 7.1 Differences in potential and kinetic energy of one ethanol surrounded by 3,000 water molecules for different cut-off's

Run	Cut-off [nm]	E_{pot} [kJ/mol]	E_{kin} [kJ/mol]
1	0.5	−85,911	33,463
2	1.4	−128,801	33,459
3	2.0	−128,968	33,458

−20.98 kJ/mol (Cabani et al. 1981). Thus, the evaluation of the parameters of any force-field, which should describe properties of solutions in an exact manner, has to consider these facts to avoid artificial effects. As $G(1)-G(2)$ is the quantity resulting from a simulation, the term to be subject of adjustment is given by:

$$G(1) - G(2) - RT\left(n_i \ln \frac{c_i}{c_o} + n_{H_2O} \ln x_{H_2O}\right) = n_i\left(\mu_i^o - \mu_i^{o,g}\right) \tag{7.33}$$

for an arbitrary species i.

Another problem arises from the so-called cut-off distance for calculating the coulomb and van der Waals interactions between sites during a MD simulation. Short cut-off distances will fasten the calculation, but lead to more inaccurate results. Using boundary conditions and the PME method, this cut-off must not exceed the half of the cubic box length. To get insight into the consequences for different values of the cut-off distance, we will analyze the total kinetic and potential energy of a system containing 1 molecule of ethanol and 3,000 molecules of water at 1 bar and 298.15 K. For this, we do three independent runs of a 1,000 ps MD simulation and calculate the mean of these energies over a time interval from 500 ps to 1,000 ps with the help of the GROMACS command `g_energy`. The result may look like given in Table 7.1:

We see, that the results corresponding to the first cut-off value differ considerably from that of run 2 and 3, which exhibit nearly the same results with respect to the computational error. A 1,000 ps MD simulation with 3,000 water molecules and one molecule of ethanol using a twin-range cut-off between 0.8 and 1.4 nm (Villa and Mark 2002) exhibits a kinetic energy of 33,975 kJ/mol and a potential energy of −130,183 kJ/mol. Both values are different to the values given in Table 7.1 and reflect the importance of the simulation conditions.

7.2.2 *Example 2: Ligand-Receptor-Complex and Affinity – Conceptual Considerations*

Now let us apply the concept, discussed so far to evaluate the equilibrium constant according to Eq. 7.3. To do so, we must have knowledge about the reference chemical potentials according to Eq. 7.11. To elucidate the application of coupling parameters let us have a look on the system state 1 consisting of n_L moles of ligand molecules and n_S moles of solvent molecules at a given pressure and temperature. We assume that n_S is much larger than n_L. Thus, the interactions between the ligand particles may be neglected just as the influence of the ligand molecules on the solvation of a particular ligand molecule. The system state 2 will be defined for all interactions between ligand and solvent particles switched off, so we have a system containing the ligand as an ideal gas and the pure solvent. Making use of the concept of the foregoing section we are able to calculate the difference $G(2){-}G(1)$. We may write the expressions for $G(1)$ and $G(2)$ from a thermodynamic point of view:

$$G(1) = n_L \cdot \left(\mu_L^o + RT \ln \left(\frac{c_L}{c_o} \right) + n_S \cdot \left(\mu_s^* + RT \ln (x_s) \right) \right). \tag{7.34}$$

Here, again for the solvent, we make use of the reference chemical potential for the pure state of water at 1 bar and 298.15 K.

$$G(2) = n_L \mu_L^{o,g} + n_s \mu_s^* \tag{7.35}$$

where μ_s^* denote the reference chemical potential of the pure solvent with mole fraction x_S and $\mu_L^{o,g}$ denotes the reference chemical potential of the ligand as an ideal gas.

Substracting $G(1)$ from $G(2)$ yields:

$$\begin{aligned} G(2) - G(1) = n_L &\left(\mu_L^{o,g} - \mu_L^o - RT \ln \left(\frac{c_L}{c_o} \right) \right) \\ &+ n_s \left(\mu_s^* - \mu_s^* - RT \ln (x_s) \right). \end{aligned} \tag{7.36}$$

As the terms for the reference potential of the solvent cancel the Eq. (7.36) now reads:

$$G(2) - G(1) = n_L \left(\mu_L^{o,g} - \mu_L^o - RT \ln \left(\frac{c_L}{c_o} \right) \right) - n_S RT \ln (x_S). \tag{7.37}$$

Next we will do a similar calculation for the ligand-receptor-complex where the interactions between the ligand and the receptor will be switched off. We define the starting state 3 composed of n_{LR} moles of ligand-receptor-complexes and n_S moles of solvent where n_S is much larger than n_{LR} and n_L. By switching of the interactions between the ligand and the receptor, we arrive at state 4 comprising of the ligand as an ideal gas and the empty receptor located in the solvent. The Gibbs energy of state 3, *G(3)*, and state 4, *G(4)*, are represented by the Eqs. 7.38 and 7.39:

$$G(3) = n_{LR} \left(\mu_{LR}^o + RT \ln \left(\frac{c_{LR}}{c_o} \right) \right) + n_s \left(\mu_s^* + RT \ln (x_s) \right) \tag{7.38}$$

$$G(4) = n_L \mu_L^{o,g} + n_R \left(\mu_R^o + RT \ln \left(\frac{c_R}{c_o} \right) \right) + n_s \left(\mu_s^* + RT \ln (x_s) \right). \tag{7.39}$$

Formulating the difference $G(4){-}G(3)$, the entire solvent terms cancel and we get:

$$\begin{aligned} G(4) - G(3) = n_L \mu_L^{o,g} &+ n_R \left(\mu_R^o + RT \ln \left(\frac{c_R}{c_o} \right) \right) \\ &- n_{LR} \left(\mu_{LR}^o + RT \ln \left(\frac{c_{LR}}{c_o} \right) \right). \end{aligned} \tag{7.40}$$

Because the ligand, the receptor and the ligand-receptor-complex are charged generally, appropriate counter ions have to be present in an electrically neutral solution. For the discussion of the thermodynamics of the association process, we presuppose, that these counter ions do not influence the formation of the ligand-receptor-complex.

For the process under consideration we have

$$n_{LR} = n_L = n_R. \tag{7.41}$$

So, the concentration terms cancel and the difference yields:

$$G(4) - G(3) = n_L \left(\mu_L^{o,g} + \mu_R^o - \mu_{LR}^o\right). \tag{7.42}$$

Establishing the difference:

$$(G(2) - G(1)) - (G(4) - G(3)) \tag{7.43}$$

will yield the following equation:

$$\begin{aligned}(G(2) - G(1)) - (G(4) - G(3)) = n_L \left(\mu_{LR}^o - \mu_R^o - \mu_L^o\right)\\ - n_L RT \ln \frac{c_L}{c_o} - n_S RT \ln x_S\end{aligned} \tag{7.44}$$

because the terms containing $\mu_L^{o,g}$ cancel. The first expression within parenthesis on the right hand side of Eq. 7.44 equals the desired quantity ΔG° in the case of $n_L = 1$ mol. The second and third term represent the corrections analogous to the solvation problem, discussed in this chapter. Applying the concept of the coupling parameter within the framework of MD simulations, we are able to evaluate the equilibrium constant for the association process, Eq. 7.1, according to Eq. 7.24.

7.2.3 *Example 1: Ethanol in Water – Practical Considerations*

After discussing the theoretical concept in the field of thermodynamics in combination with thermodynamic integration method (coupling-parameter method), we will present the calculation of the Gibbs energy of solvation using the software package GROMACS. Let us exemplary calculate the Gibbs energy of solvation for ethanol in water. Therefore, the following steps have to be performed.

- ☐ Construct the molecule, for which the Gibbs energy of solvation should be calculated and save the coordinates as pdb-file
- ☐ Contact the PRODRG-Server in order to obtain the GROMACS-coordinates as a *gro*-file and the GROMACS-topology as a *top*-file (see Chap. 4)
- ☐ Change the size of the simulation box in an appropriate manner
- ☐ Minimize your molecule in gas phase (see Chap. 6)

- ☐ Calculate $^{dG}/_{d\lambda}$ for gas phase via MD simulation, for example via the shell script `gibbs_energy` (see below)
- ☐ Center your solute, obtained by item 4 in the simulation box
- ☐ Solvate your solute with the desired solvens using the GROMACS command `genbox`
- ☐ Adopt your *mdp*-file for minimization, if necessary
- ☐ Minimize your simulation box (solute in solvens)
- ☐ Adopt your *mdp*-file for molecular dynamic simulation, if necessary
- ☐ Calculate $^{dG}/_{d\lambda}$ for ethanol in water via MD simulation, for example via the shell script `gibbs_energy` (see below)

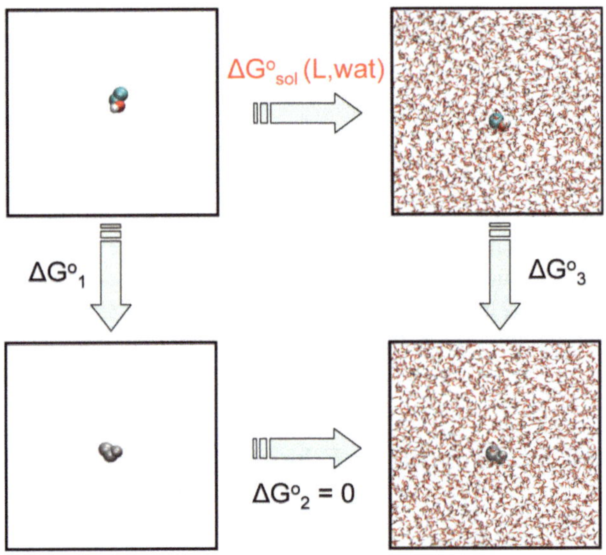

Fig. 7.1 Thermodynamic cycle for ethanol in water (*coloured ligand*: full interactions; *grey ligand*: no Coulomb and van der Waals interactions)

Each of the mentioned items will be explained now step by step: First, generate an appropriate *gro*- and *top*-file for ethanol, as described in detail in Chap. 4. Save the files, shown explicitly in Chap. 4, as `ethanol.gro` and `ethanol.itp`. In the next step, one has to modify the topology file `ethanol.itp`. In the section `[ atoms ]`, only the atom types, charges and masses for the state 1 are defined. To calculate the Gibbs energy of solvation according to the thermodynamic cycle (Fig. 7.1; Villa and Mark 2002), the solute has to be transferred into an ideal gas state by switching of all the Coulomb and van der Waals interactions within ethanol.

The energy, which is necessary to remove all non-bonded internal interactions, like Coulomb- or van der Waals interaction for the solute in vacuum is represented by ΔG_1^o (Fig. 7.1). Therefore, all sites of state 1 are mutated gradually into dummy sites, corresponding to state 2. In GROMACS, for example, a dummy is represented by DUM, where all partial charges and van der Waals interaction parameters are set to zero. Therefore, in the section `[ atoms ]` three additional columns for type, charge and mass of the dummy sites are necessary. Setting the charges of all solute sites to zero and defining the site type DUM, switches off the Coulomb and van der Waals interactions between the sites of the solute. However, all internal, bonded interactions and masses remain unchanged. The analogous energy term to remove all non-bonded internal interactions for the solute and between the solute and the solvent is described by ΔG_3^o (Fig. 7.1), neglecting the concentration terms for ethanol and water as discussed in this chapter in the framework of parameter adjustment. Therefore, the same procedure as mentioned above for ΔG_1^o has to be performed. The energy, which is necessary to transfer the dummy solute from vacuum into solvent is described by ΔG_2^o (Fig. 7.1). Since there is no interaction of the dummy solute with the remaining system, this value is zero. The values for ΔG_1^o and ΔG_3^o can be calculated by appropriate MD simulations, as shown later. Subsequently, the Gibbs energy of solvation of ethanol in water $\Delta G_{sol}^o(EtOH)$ can be estimated via

$$\Delta G_{sol}^o(EtOH) = \Delta G_1^o - \Delta G_3^o. \tag{7.45}$$

In the following, the `[ atoms ]` section of `ethanol.itp` with the additional column is presented.

```
[ atoms ]
;  nr   type  resnr resid atom  cgnr  charge mass
   1    CH3     1   DRG    CAA   1     0.000  15.0350  DUM 0.0 15.0350
   2    CH2     1   DRG    CAC   1     0.266  14.0270  DUM 0.0 14.0270
   3     OA     1   DRG    OAB   1    -0.674  15.9994  DUM 0.0 15.9994
   4     HO     1   DRG    HAA   1     0.408   1.0080  DUM 0.0 1.0080
```

Save this file as `ethanol.itp`. Minimize the ethanol and save the resulting file as *ethanol_gas.gro*. Solvate the minimized ethanol with an appropriate number of water molecules, using the GROMACS commands `editconf` to center the solvent in the simulation box and `genbox` for solvation. Minimize the box and save the file as `ethanol_sol.gro`. Now, one can start the simulation to calculate the Gibbs energy of solvation. Therefore, you would have to start a distinct number of subsequent molecular dynamic simulations for discrete values of λ. Thereby, it should be taken into account, to perform the calculations at least at eighteen values for λ, for example: 0.0, 0.05, 0.1, 0.2, 0.3, 0.4, 0.45, 0.5, 0.55, 0.6, 0.7, 0.8, 0.9, 0.95, 0.975, 0.99, 0.995 and 1.0. In general, a $\Delta\lambda$ of 0.1 can be used. But to avoid singularities around 0, 0.5 and 1, at these regions, smaller $\Delta\lambda$ are recommended, as shown above. It is very useful to perform the calculations for aqueous and gaseous phase as in two different subdirectories `aqueous` and `gaseous`. Additionally, for each λ-calculation a separate subdirectory in the subdirectory `aqueous` or `gaseous` should be used. Surely, you can perform the simulations for each λ manually, but this is very inefficient. Therefore, it is more useful to use the shell script `gibbs_energy`, shown below. Using this script, all simulations of the `aqueous` or `gaseous` phase can be performed.

```
#!/bin/tcsh -f

set n = 18

set base = lambda

set i = 1

set lambdas = (0.0 0.05 0.1 0.2 0.3 0.4 0.45 0.5 0.55 0.6 0.7 0.8 0.9 0.95 0.975 0.99 0.995 1.0)

while ($i <= $n)
  mkdir ${base}_${i}
  cd ${base}_${i}
  cp ../ethanol.top .

if ($i == 1) then
  cp ../system_min.gro .
  cp ../md_first.mdp .
  grompp -f md_first -o md_first -c system_min -r system_min -p ethanol
  wait
  mdrun -v -s md_first -e md_first -o md_first -c after_md -g shortlog
  wait
else
  head -n 65 ../md.mdp> ./md.mdp
  echo "init_lambda=$lambdas[$i]" >>./md.mdp
  @ k = $i - 1
  cp ../${base}_$k/after_md.gro ./system_min.gro
  grompp -f md -o md -c system_min -r system_min -p ethanol
  wait
  mdrun -v -s md -e md -o md -c after_md -g shortlog
  wait
endif

cd ..
@ i++

end

set i = 1

```

```
41 while ( $i <= 18 )
42   tail -n 100001 lambda_${i}/dgdl.xvg
   > lambda_${i}/dgdl.dat
43 echo "$i `average_gibbs lambda_${i}/dgdl.dat` "
   >> lambda_gibbs.dat
44 @ i++
45 end
```

Before the script gibbs_energy can be started, the execute permission for the user has to be set using the following command:

```
> chmod u+x gibbs_energy ↵
```

Following, an exemplary mdp-file, named md_first.mdp for calculation in aqueous phase is given. We like to suggest explicitly, that the simulation parameters may be adopted by the user as appropriate for the simulation problem. In the mdp-file shown below, the parameter init_lambda is set to zero for the first calculation with full interactions. The file *md.mdp* is identical with the file md_first.mdp, only the parameter unconstrained_start is set to yes and gen_vel is set to no. In the script gibbs_energy (see above), line 24 and 25, the last line of the file md.mdp is adopted with regard to the actual value of λ in the parameter init_lambda (line 66 in md.mdp). For calculation in gaseous phase, the *mdp*-files have to be adopted in an appropriate manner.

```
 1;          md_first.mdp
 2;          MD
 3;
 4;          Input file
 5;
 6 title                       =Ethanol in water
 7 cpp                         =/lib/cpp
 8 ;define                     =-DPOSRES
 9 ;constraints                =all-bonds
10 ;constraint_algorithm       =lincs
11 unconstrained_start         =yes
12 integrator                  =md
13 tinit                       =0
14 dt                          =0.001; ps!
15 nsteps                      =100000
16 nstcomm                     =1
17 ;Output control
18 nstxout                     =5000
19 nstvout                     =5000
20 nstfout                     =0
21 nstlog                      =5000
22 nstenergy                   =100
```

```
;Neighbor searching
nstlist                 =10
ns_type                 =grid
pbc                     =xyz
rlist                   =1.4
;Electrostatics and VdW
coulombtype             =PME
;rcoulomb_switch        =0
rcoulomb                =1.4
epsilon_r               =1.0
;epsilon_rf             =7.0
vdwtype                 =Cut-off
;rvdw_switch            =0
rvdw                    =1.4
;DispCorr               =EnerPres
fourierspacing          =0.12
fourier_nx              =0
fourier_ny              =0
fourier_nz              =0
pme_order               =4
ewald_rtol              =1e-5
optimize_fft            =yes
; Temperature coupling
tcoupl                  =berendsen
tc-grps                 =LIG SOL
tau_t                   =0.1 0.1
ref_t                   =298 298
; Energy monitoring
energygrps              =LIG SOL
; Pressure coupling is not on
Pcoupl                  =berendsen
pcoupltype              =isotropic
tau_p                   =0.5 0.5 0.5 0.0 0.0 0.0
compressibility         =4.5e-5 4.5e-5 4.5e-5 0.0
                         0.0 0.0
ref_p                   =1.0
; Generate velocites is on at 298 K.
gen_vel                 =yes
gen_temp                =298
gen_seed                =173529
; Free Energy Calculation
free_energy             =yes
sc-alpha                =1.5
sc-power                =2
init_lambda             =0.0
```

The files `dgdl.dat`, created by `gibbs_energy` contain two columns. The first column represents the time and the second one the values $^{dG}/_{d\lambda}$. To calculate the mean of $^{dG}/_{d\lambda}$ at one λ, the gawk-script `average_gibbs`, presented below, can be used:

```
#!/usr/bin/gawk -f

BEGIN {s=0; n=0}

{n++; s=s+$2}

END {print s/n}
```

Before the script `average_gibbs` can be started, the execute permission for the user has to be set using the following command:

```
>chmod u+x average_gibbs ↵
```

Additionally, `average_gibbs` must reside in the same directory, as `gibbs_energy`, because `average_gibbs` is started in line 43 of `gibbs_energy`. Thus, after `gibbs_energy` has completed, a file named `lambda_gibbs.dat` is created, containing λ in the first column and the mean of $^{dG}/_{d\lambda}$ in the second column. With the help of the script `integrate`, shown later on, the integration can be performed.

The calculation for ethanol in water, ethanol in vaccuo and ethanol for the transfer from vaccuo into aqueous phase, may lead to $^{dG}/_{d\lambda}$ as function of λ at a temperature of 293.0 K as shown in Figs. 7.2, 7.3 and 7.4.

To be able to calculate the Gibbs energy of solvation, the integral for the corresponding curve has to be determined. For integration you can use distinct software products, like `xmgrace` (http://plasma-gate.weizmann.ac.il/Grace/). However, after long simulations, the data sets might be too large and it might be very time consuming to perform the integration with such software. In this case, you can use your own C-code or `gawk`. Thus, in the following it is shown, how to write an integration routine by yourself. In literature, a large number of numeric methods with regard to numeric integration are described, like the Simpson's rule or the trapezoidal rule. A

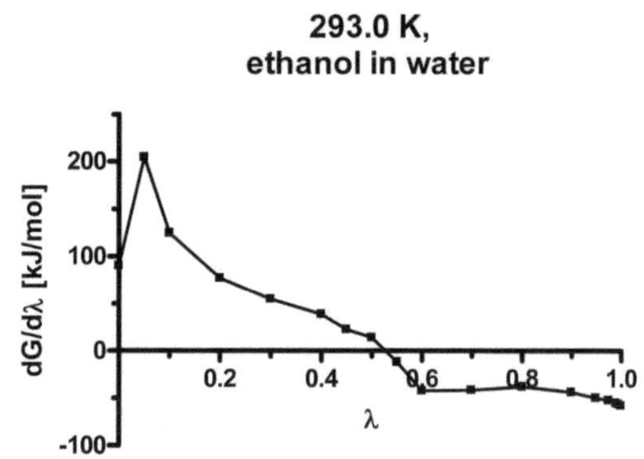

Fig. 7.2 $^{dG}/_{d\lambda}$ as function of the coupling parameter λ for ethanol in water

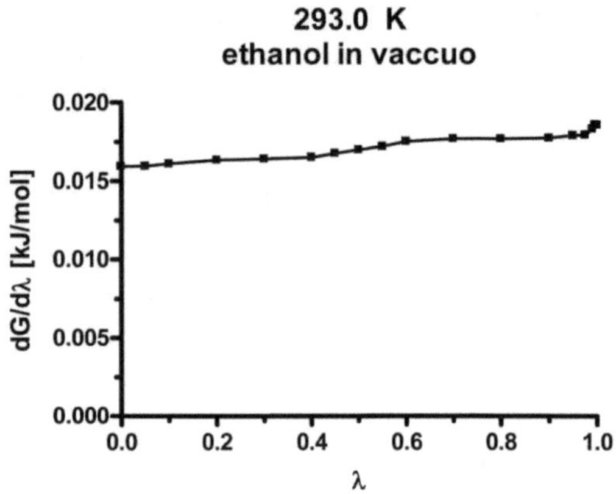

Fig. 7.3 $^{dG}/_{d\lambda}$ as function of the coupling parameter λ for ethanol in vaccum

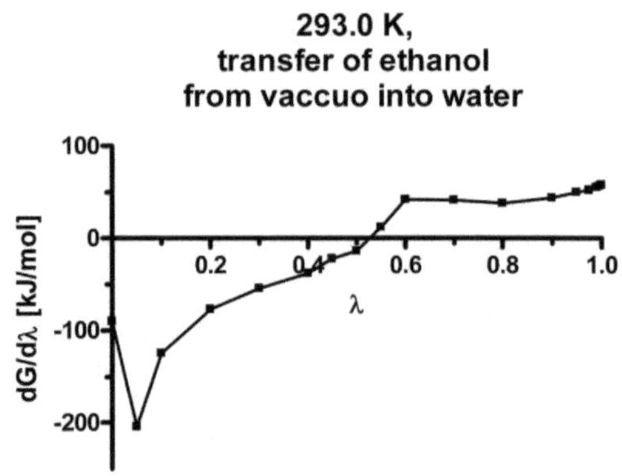

Fig. 7.4 $^{dG}/_{d\lambda}$ as function of the coupling parameter λ for the transfer of ethanol from vaccuo into water; this course was calculated based on the data presented in Figs. 7.2 and 7.3

very simple and stable numeric method is the trapezoidal rule, and thus, we focus onto this.

The formula for the trapezoidal rule is given by Eq. 7.46:

$$\int_{x_o}^{x_o+h} f(x)dx \approx \frac{h}{2} \cdot (f(x_o) + f(x_o + h)). \tag{7.46}$$

Via summation over all intervals, Eq. 7.47 is obtained:

$$\int_{x_o}^{x_n} f(x)dx \approx h \cdot \left[\frac{f(x_o)}{2} + f(x_1) + f(x_2) + \ldots + f(x_{n-1}) + \frac{f(x_n)}{2}\right]. \tag{7.47}$$

Be aware, that the equation, mentioned above, is only valid, if n intervals of equal width are used! In all other cases, a modified formula has to be used:

$$\int_{x_1}^{x_n} f(x)dx \approx \frac{1}{2}\sum_{i=1}^{n-1} ((x_{i+1} - x_i)\,(f(x_{i+1}) + f(x_i))). \tag{7.48}$$

Table 7.2 Derivative of the Gibbs energy with respect to the coupling parameter for the transfer of ethanol from vacuum into water at 293.0 K

λ	0.00	0.05	0.10	0.20	0.30	0.40
dG/dλ [kJ/mol]	−90.02	−204.43	−124.34	−76.86	−54.48	−38.24
λ	0.45	0.50	0.55	0.60	0.70	0.80
dG/dλ [kJ/mol]	−22.34	−13.78	11.96	41.98	41.61	37.91
λ	0.90	0.95	0.975	0.99	0.995	1.00
dG/dλ [kJ/mol]	43.95	50.02	52.40	55.31	56.47	57.90

Therein, x_0 corresponds to the lower bound of integration and x_n to the upper bound of integration

Now, let us apply this integration method to the example ethanol in water, shown above, in order to calculate the solvation energy. The data, presented in Fig. 7.4, are given in Table 7.2 and represent the derivative of the Gibbs energy with respect to λ as a function of λ.

The integration can easily be performed using Eq. 7.48 with the following gawk-script, named `integrate`:

```
#!/usr/bin/gawk -f

BEGIN {s=0; n=0; OFS=" "}

{n++; x[n]=$1; y[n]=$2}

END {for(i=1;i<n;i++)
s+=0.5*(x[i+1]−x[i])*(y[i+1]+y[i]);
print "dG_solv=",s," kJ/mol"}
```

Now, you can open an editor, write the command sequences into the editor and save the file with the name `integrate`. To test this script, you should first change your file access rights, using the following command:

```
>chmod u+x integrate ↵
```

Next, create a file, containing the data shown in Table 7.2. In this example, we name the file `ethanol_sol.dat`, which should look like this:

```
0.00 −90.02
0.05 −204.43
0.10 −124.34
0.20 −76.86
0.30 −54.48
0.40 −38.24
0.45 −22.34
0.50 −13.78
0.55 11.96
0.60 41.98
0.70 41.61
```

```
0.80 37.91
0.90 43.95
0.95 50.02
0.975 52.40
0.99 55.31
0.995 56.47
1.00 57.90
```

For integration, perform the following command:

```
>cat ethanol_sol.dat | integrate ↵
```

The output should like this:

```
dG_solv=-20.7068 kJ/mol
```

Sometimes it might be the case, that you have no pure `dat`-file, containing the two columns, as given in `ethanol_sol.dat`. Perhaps you have a `xvg`-file – `xvg`-files are often an output of GROMACS. Here you see an example-file `ethanol_sol.xvg`.

```
# This file was created Thu Dec 11 11:43:19 2008
# by the following command:
# mdrun -v -s md -e md -o md -c after_md -g shortlog
#
# mdrun is part of G R O M A C S:
#
# GROwing Monsters And Cloning Shrimps
#
@ title "dG/d\81\4"
@ xaxis label "lambda"
@ yaxis label "dG/d\81\4 (kJ mol\S-1\N[\81\4]\S-1\N)"
@ TYPE xy
0.00 -90.02
0.05 -204.43
0.10 -124.34
0.20 -76.86
0.30 -54.48
0.40 -38.24
0.45 -22.34
0.50 -13.78
0.55 11.96
0.60 41.98
0.70 41.61
0.80 37.91
0.90 43.95
0.95 50.02
0.975 52.40
```

```
0.99 55.31
0.995 56.47
1.00 57.90
```

In this case, you cannot use the command shown above. Here, you have two alternatives: First, you delete all lines, except the data lines. Or secondly, and more elegant

```
>grep -v '^[#@]' ethanol_sol.xvg|integrate ↵
```

What does this command do? If you take a closer look onto the file `ethanol_sol.xvg`, shown above, you see, that additionally to the data lines, there are lines starting with the symbol # or @. These lines have to be deleted, which can be done with the command

```
>grep -v '^[#@]' ethanol_sol.xvg ↵
```

The option `-v` inverts `grep`'s search: all lines, not containing one of the characters # or @ in the specified pattern at the beginning of the line, indicated by ^, will be printed and may be used as input to the command `integrate` (see Chap. 11).

Additionally, calculations of Gibbs energy of solvation can be performed at different temperatures. This allows to calculate enthalpy and entropy of solvation. You can do so for example with ethanol in water. In Table 7.3, the predicted temperature dependence of the Gibbs energy of solvation of ethanol in water is shown.

Table 7.3 Predicted values for the Gibbs energy of solvation ΔG^o_{sol} at different temperatures for ethanol in water

T [K]	ΔG^o_{sol} [kJ/mol]
283	-21.1 ± 0.3
288	-21.0 ± 0.4
293	-20.7 ± 0.2
298	-20.5 ± 0.4
303	-20.5 ± 0.3

For calculation of enthalpy and entropy of solvation, the following Eq. 7.49 can be used for linear fit to obtain: ΔH^o_{sol}, ΔS^o_{sol}, and $\Delta C^o_{p,sol}$:

$$\Delta G^o_{sol}(p,T) = \Delta H^o_{sol}(p,T_o) + \Delta C^o_{p,sol} \cdot (T - T_o) - T \cdot \left(\Delta S^o_{sol}\,(p,T_o) + \Delta C^o_{p,sol} \cdot \ln\left(\frac{T}{T_o}\right)\right). \tag{7.49}$$

Usually, the fit can be performed with adequate software. Besides that, you can program a leastsquare method by yourself. T_o was set to 293 K. After fitting, the following thermodynamic parameters were obtained for a temperature of 293 K (Table 7.4):

Table 7.4 Predicted thermodynamic reference parameters at 293 K for the solvation of ethanol in water

ΔG^o_{sol}	-20.7 ± 0.2 kJ/mol
ΔH^o_{sol}	-30.6 ± 1.7 kJ/mol
ΔS^o_{sol}	-33.7 ± 5.8 J/(mol K)
$\Delta C^o_{p,sol}$	498 ± 572 J/(mol K)

The data points and the corresponding fit are shown in Fig. 7.5.

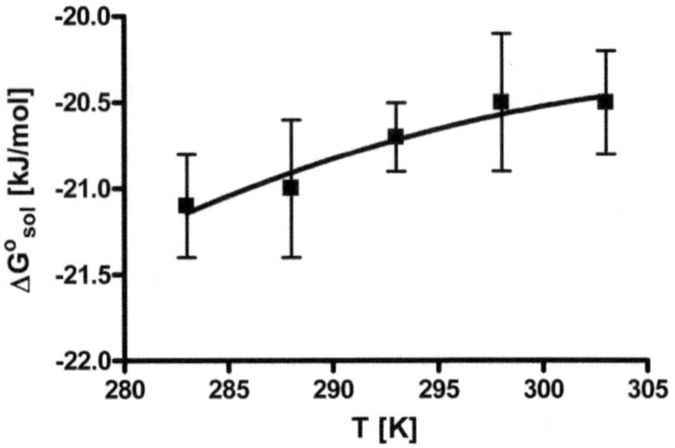

Fig. 7.5 Predicted values for the Gibbs energy of solvation for ethanol at different temperatures

Alternatively to the method presented above, method 1, for calculation of the Gibbs energy of solvation, a second method – method 2 – can be used: To transfer gaseous ethanol from ideal gas state at 1 bar and an arbitrary temperature into pure solvent, to obtain an ideal solution of ethanol in water, which corresponds to the difference $G(1){-}G(2)$ (cf. Eq. 7.37) the following mdp-file has to be applied:

```
;
;           MD
;
;           Input file
;
title                        =Ethanol in water
cpp                          =/lib/cpp
;define                      =-DPOSRES
;constraints                 =all-bonds
;constraint_algorithm        =lincs
unconstrained_start          =yes ; or no, as appropriate
integrator                   =sd1
tinit                        =0
dt                           =0.001  ; ps!
nsteps                       =1000000
nstcomm                      =1
; Output control
nstxout                      =1000
nstvout                      =1000
nstfout                      =0
nstlog                       =1000
nstenergy                    =1000
; Neighbor searching
```

```
nstlist                  =10
ns_type                  =grid
pbc                      =xyz
rlist                    =1.4
; Electrostatics and VdW
coulombtype              =pme
;rcoulomb_switch         =0
rcoulomb                 =1.4
epsilon_r                =1.0
;epsilon_rf              =7.0
vdwtype                  =Cut-off
;rvdw_switch             =0
rvdw                     =1.4
;DispCorr                =EnerPres
fourierspacing           =0.135
fourier_nx               =0
fourier_ny               =0
fourier_nz               =0
pme_order                =4
ewald_rtol               =1e-5
ewald_geometry           =3dc
optimize_fft             =yes
; Temperature coupling
tcoupl                   =berendsen
tc-grps                  =system
tau_t                    =0.1
ref_t                    =298.15
; Energy monitoring
energygrps               =system
; Pressure coupling is on
Pcoupl                   =berendsen
pcoupltype               =isotropic
tau_p                    =0.5 0.5 0.5 0.0 0.0 0.0
compressibility          =4.5e-5 4.5e-5 4.5e-5 0.0 0.0
                          0.0
ref_p                    =1.0
; Generate velocites is on at 298.15 K.
gen_vel                  =yes
gen_temp                 =298.15
gen_seed                 =173529
;
free_energy              = yes
init_lambda              = 0
delta_lambda             = 0.000001
```

```
sc_alpha          = 1.5
sc_power          = 2.0
couple-moltype    = EtOH
couple-lambda0    = vdw-q
couple-lambda1    = none
couple-intramol   = no
```

The last nine lines of this file govern the calculation of ΔG^o_{sol}. For an explanation of the keywords and the corresponding values, the reader is referred to the GROMACS manual (van der Spoel et al. 2005). The comparison of the integration cycles according to method 1 and 2, presented in Fig. 7.6 shows a good accordance. The Gibbs energy of solvation of ethanol, using method 2 is predicted to be (-24.3 ± 1.7) kJ/mol.

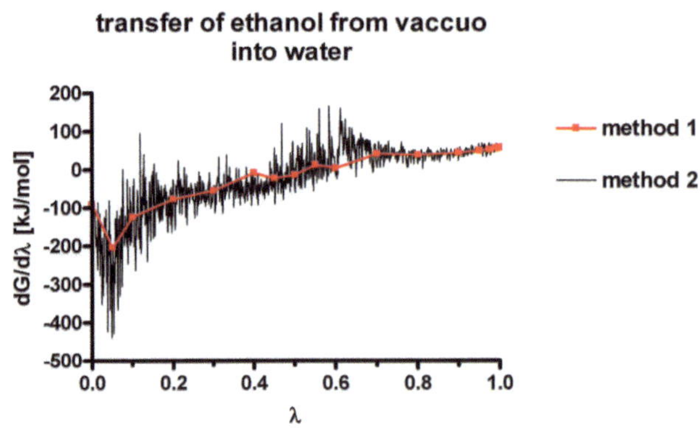

Fig. 7.6 Comparison of $^{dG}/_{d\lambda}$ in dependence of the coupling parameter λ for the transfer of ethanol from vaccuo into water at a temperature of 298.15 K calculated with method 1 and method 2

Experimental data In general, it is very important to compare properties, predicted by molecular modelling techniques, with experimental data. This is also recommended for predicted thermodynamic parameters, like ΔG^o_{sol}, ΔH^o_{sol} or ΔS^o_{sol}. A large number of thermodynamic parameters of solvation can be found in literature (Cabani et al. 1981; Abraham 1984). Such comparison of predicted data with known experimental data is necessary to judge the predictive quality of a molecular modelling technique and/or the used force field parameters.

The predicted ΔG^o_{sol} value of ethanol (Table 7.4) is in very good accordance with the experimentally determined value (Table 7.5). In contrast, the difference between prediction (Table 7.4) and experiment (Table 7.5) for ΔH^o_{sol} and ΔS^o_{sol} is larger, than for ΔG^o_{sol}. A reason for this difference might be, that the force field parameters were optimized only with regard to ΔG^o_{sol}, but not with regard to ΔH^o_{sol} and ΔS^o_{sol} (Villa and Mark 2002). A more detailed comparison between predicted and experimental ΔG^o_{sol} values for analogues of amino acid side chains can be found in literature (Villa and Mark 2002).

Table 7.5 Thermodynamic parameters of solvation for ethanol at 25 °C (Cabani et al. 1981). The standard entropy of solvation was calculated based on the standard Gibbs energy and enthalpy of solvation

ΔG^o_{sol}	−20.98 kJ/mol
ΔH^o_{sol}	−52.40 kJ/mol
ΔS^o_{sol}	−105.4 J/(mol K)

7.2.4 Example 2: Gibbs Energy of Binding

Above, we introduced the calculation of the Gibbs energy of solvation of a ligand in water. However, we are not interested in this value. Rather, in context of GPCRs, we are interested in the Gibbs energy of binding ΔG_{bind} of a ligand from aqueous phase into the binding pocket of a GPCR (Fig. 7.7). To calculate this quantity, the concept of thermodynamic integration, already shown for ethanol in water can be used as well.

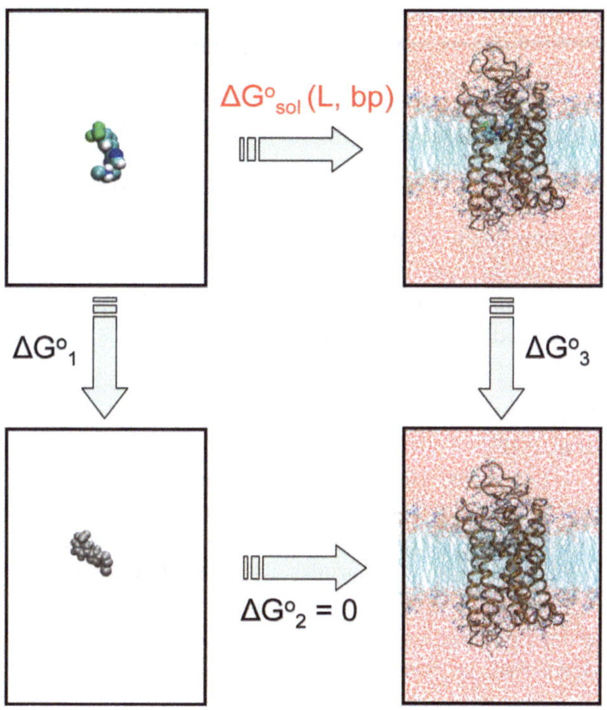

Fig. 7.7 Thermodynamic cycle for a ligand in the binding pocket of a receptor (*coloured ligand*: full interactions; *grey ligand*: no Coulomb and van der Waals interactions)

Before you can start to calculate the Gibbs energy of solvation of a ligand in the binding pocket of a GPCR, you have to dock the interesting ligand into the binding pocket and perform molecular dynamic simulations, as described in the corresponding Chap. 6 in order to obtain a stable ligand-receptor-complex. During the thermodynamic cycle, the interaction between the ligand and the surrounding is switched off. In case of a homogeneous surrounding, like water or another solvens

Fig. 7.8 Thermodyamic cycle for ligand-receptor interaction with mutation of the ligand. (Henin et al. 2006)

this does not matter. But in case of a specific location of the ligand in the binding mode, this might lead to problems: Due to the decreasing interaction between ligand and receptor, the ligand may be able to wander around somewhere in the simulation box. Consequently, the wrong surrounding will be included into the calculation. Thus, it will be very useful in a lot of cases to put slight position constraints onto the ligand, after equilibration in the binding pocket via MD simulation with full interaction.

Example 2.1 Within a study, addressing the human cholecystokinin-1 receptor, free energy calculations were used to compare predicted changes in Gibbs energy of binding (ΔG_{bind}) with respect to mutation of the ligand (Henin et al. 2006). For this purpose, the authors use the thermodynamic cycle presented in Fig. 7.8.

For calculation of the change in Gibbs energy of binding for mutation of the ligand CCK9, the authors use the following Eq. 7.50.

$$\Delta G_{bind}^{mut} - \Delta G_{bind} = \Delta G_2^{mut} - \Delta G_1^{mut}. \tag{7.50}$$

A comparison of the experimental and predicted results is given in Table 7.6. In general, the correlation between experimental and calculated data is well. Thus, this method may be useful for predicting the influence of a structural modification within the ligand with regard to binding affinity.

Example 2.2 Within a pharmacological study, the binding affinity of several phenylhistamines at the human histamine H_4 receptor (hH_4R) was determined (Wittmann et al. 2011).

The pharmacological data revealed large differences in binding affinity of the six phenylhistamines at hH_4R, as shown in Table 7.7. In order to explain the pharmacological data, molecular modelling studies were performed. Since the phenylhistamines act as partial agonists at hH_4R, these compounds were docked into an active state model and subsequently, molecular dynamic simulations were performed, in order to obtain a stable binding mode. The binding mode for two phenylhistamine-derivatives is shown in Fig. 7.9.

Table 7.6 Comparison of experimental (exp) and calculated (calc) changes in Gibbs energy of binding with respect to mutations in the ligand CCK9 (Henin et al. 2006). The terms $\Delta G_{bind}^{\exp}$ and ΔG_{bind}^{calc} represent the change in Gibbs energy of binding for structural change of the ligand CCK9

	K_i [nM]	$\Delta G_{bind}^{\exp}$ [kcal/mol]	ΔG_{bind}^{calc} [kcal/mol]
compound CCK9	1.38 ± 0.15	–	–
Asp-8 → Ala	253.8 ± 11.4	3.2 ± 0.3	3.0 ± 0.7
S-Tyr-3 → Tyr	108.8 ± 4.8	2.7 ± 0.1	1.9 ± 0.4

Table 7.7 Binding affinities (pK_i) of six phenylhistamine derivatives at hH_4R at a temperature of 298.15 K. (Wittmann et al. 2011)

		R^1	R^2	$pK_i\ (hH_4R)$
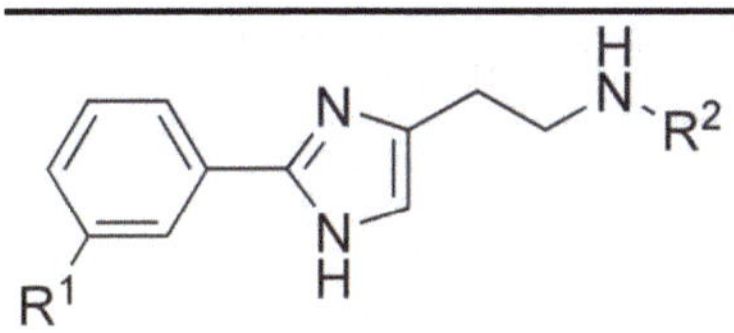	PheHIS-1	H	H	4.79 ± 0.04
	PheHIS-2	CF_3	H	5.91 ± 0.12
	PheHIS-3	Br	H	5.76 ± 0.01
	PheHIS-4	H	CH_3	6.13 ± 0.08
	PheHIS-5	CF_3	CH_3	6.80 ± 0.04
	PheHIS-6	Br	CH_3	6.56 ± 0.06

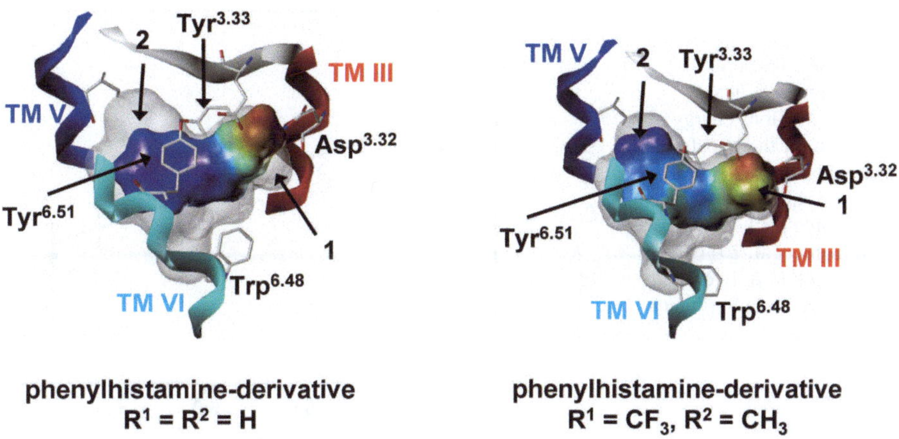

Fig. 7.9 Binding mode of two phenylhistamine derivatives PH-1 (*left*) and PH-5 (*right*) in the binding pocket of hH_4R. (Wittmann et al. 2011, copyright by Springer, with permission from Springer)

In case, that the small phenylhistamine ($R^1 = R^2 = H$) is bound to the receptor, two small empty pockets (Fig. 7.9, left: arrow 1 and arrow 2) were identified. If a more bulkier phenylhistamine ($R^1 = CF_3$, $R^2 = CH_3$) is bound to the receptor, the methyl moiety (CH_3) fits well into pocket 1 (Fig. 7.9, right: arrow 1) and the trifluoromethyl moiety (CF_3) fits well into pocket 2 (Fig. 7.9, right: arrow 2). Thus, it can be suggested that the additional methyl and trifluoromethyl moieties result in an increase of interaction between the hH_4R and ligand. This is in good accordance to higher affinity of PH-5 compared to PH-1 (Table 7.7). However, using the thermodynamic integration method, this qualitative explanation could be quantified (Table 7.8).

A correlation of the experimentally determined pK_i values with the predicted changes in Gibbs energy of solvation for the transfer of the ligands from aqueous phase into the binding pocket of hH_4R is presented in Fig. 7.10.

As revealed by Fig. 7.10, the correlation between predicted and experimental data is quite well. In this case, there is rather no difference, if the predicted $\Delta\Delta G^o_{sol}$ value for the transfer of the ligand from aqueous phase in to binding pocket of hH_4R, or the predicted ΔG^o_{sol} value of the ligand in the binding pocket of hH_4R is correlated with the experimentally determined pK_i value. However, the more accurate way would be

Table 7.8 Calculated Gibbs energies of solvation for phenylhistamines in water, in the binding pocket of hH_4R and for the transfer of aqueous phase into the binding pocket of hH_4R at a temperature of 298.15 K. (Wittmann et al. 2011)

	$\Delta G^o_{sol}(L, wat)$ [kJ/mol]	$\Delta G^o_{sol}(L, hH_4R)$ [kJ/mol]	$\Delta\Delta G^o_{sol}(L, wat \rightarrow hH_4R)$ [kJ/mol]
PH-1	-204 ± 1	-477 ± 15	-273 ± 16
PH-2	-224 ± 2	-525 ± 11	-301 ± 13
PH-3	-215 ± 2	-516 ± 10	-301 ± 12
PH-4	-190 ± 2	-512 ± 19	-322 ± 21
PH-5	-205 ± 3	-544 ± 13	-340 ± 16
PH-6	-202 ± 2	-517 ± 19	-315 ± 21

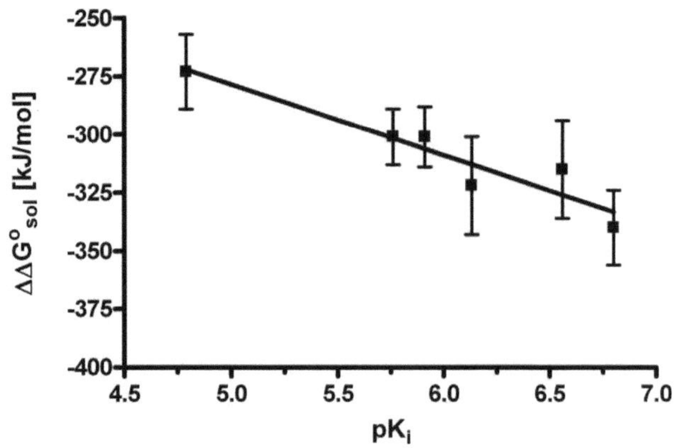

Fig. 7.10 Correlation between the predicted changes in Gibbs energy of solvation for the transfer of phenylhistamine derivatives from the aqueous phase into the binding pocket of hH_4R. (Wittmann et al. 2011, copyright by Springer, with permission from Springer)

Table 7.9 Affinities (pK_i values) of three selected compounds at hH_1R at room temperature. (Wagner et al. 2011)

	pK_i
L1	6.77 ± 0.05
L2	8.15 ± 0.10
L3 (R/S)	6.67 ± 0.09

to use the $\Delta\Delta G^o_{sol}$ value for transfer of ligand from aqueous phase into the binding pocket, because this is exactly the process, which is experimentally determined. Furthermore, one should take into account that there might be systems, where only using $\Delta\Delta G^o_{sol}$ will lead to a well correlation.

Example 2.3 A procedure, analogue to example 2.2 just above, was performed within another study addressing the histamine H_1 receptor (Wagner et al. 2011). Within this study, the affinities of selected ligands (Fig. 7.11) to human histamine H_1 receptor were determined (Table 7.9).

Having a look onto the Table 7.9, the affinity of ligand L2 compared to those of ligand L1 and L3 is significantly higher. The corresponding values of the Gibbs energy for the ligand-receptor binding process ($\Delta\Delta G^o_{sol}(L, wat \rightarrow hH_1R)$) from

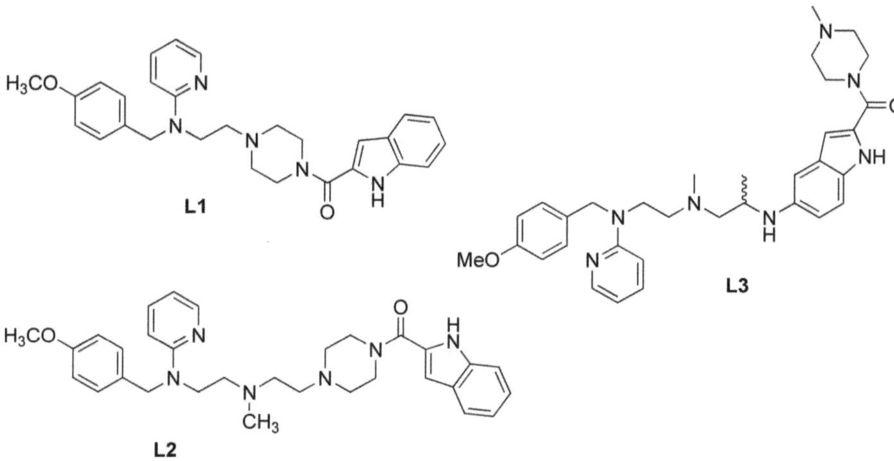

Fig. 7.11 Structures of selected compounds

Table 7.10 Calculated Gibbs energies of solvation for selected compounds (Fig. 7.11) in water, in the binding pocket of hH_1R and for the transfer of aqueous phase into the binding pocket of hH_1R at a temperature of 298.15 K. (Wagner et al. 2011, copyright by Springer, with permission from Springer)

	$\Delta G^o_{sol}(L, wat)$ [kJ/mol]	$\Delta G^o_{sol}(L, hH_1R)$ [kJ/mol]	$\Delta\Delta G^o_{sol}(L, wat \rightarrow hH_1R)$ [kJ/mol]
L1	-171 ± 2	-446 ± 21	-275 ± 23
L2	-145 ± 3	-436 ± 16	-291 ± 19
L3(*R*)	-248 ± 4	-515 ± 18	-267 ± 22
L3(*S*)	-243 ± 3	-507 ± 16	-264 ± 19

Table 7.10 are in very good accordance to the trend of the experimental data. Obviously it would be a much simpler task, only to compute the transfer of the ligand from the gaseous state into the binding pocket of the receptor ($\Delta G^o_{sol}(L, hH_1R)$) (Table 7.10; Wagner et al. 2011). But these data do not reveal the mentioned trend of affinities. Omitting the desolvation process for the ligand would pretend a higher affinity for L3(R/S) than for L2.

Chapter 8
Special Topics in GPCR Research

8.1 Interaction Between a GPCR and the Gα-subunit

It is well known that GPCRs couple in the intracellular part with G proteins, which consist of a Gα, Gβ and Gγ subunit. Induced by activation of the GPCR by an agonist the G proteins act as intracellular switches on molecular level turning on intracellular signal cascades. What is known about the interaction of a GPCR with a G protein on molecular level? On the on hand, there is an increasing number of crystal structures of GPCRs described in literature (see Appendix Important Crystal Structures of GPCRs (Source: http://www.pdb.org)) and on the other hand, some crystal structures of heterotrimeric G proteins are known. Recently, the structure of opsin, cocrystallized with a part of the C-terminus of Gα was published (Scheerer et al. 2008). But until 2011, there exists no crystal structure of a complete GPCR-G protein complex. However, within a small number of experimental and theoretical studies, the interactions between GPCR and G protein were investigated (Fanelli et al. 1999; Greasley et al. 2001; Oliveira et al. 2003; Chou 2005; Raimondi et al. 2008). In general, there is only little knowledge about the interactions between GPCR and G protein on molecular level.

In literature, two different models with regard to GPCR-G protein coupling are discussed. This is on the one hand the so called "collision coupling" model. Within this model, it is suggested that only the active receptor interacts with the G protein (Tolkovsky et al. 1978). In contrast, the second model, the so called "precoupling model" suggests that the G proteins interact with the GPCR before the receptor is activated by an (partial) agonist. Thus, GPCR and G protein are pre-coupled. The "precoupling model" is provided by several studies (Alves et al. 2003, 2005; Gales et al. 2006).

By experimental studies, some regions of GPCR and G protein, which interact with each other, were identified. In general it is supposed that a pocket in the intracellular part of the GPCR is opened during activation. And this pocket is suggested to interact with the C-terminus of the Gα subunit (Scheerer et al. 2008; Rasmussen et al. 2011). Furthermore, mutagenesis studies suggest that amino acids of the α4-β6 loop (Bae et al. 1999; Cai et al. 2001) and α3-α5 loop (Grishina et al. 2000) interact with the GPCR. Thus, by experimental studies, some important suggestions with regard to GPCR-G protein interactions could be obtained. However, there occurs

A. Strasser, H.-J. Wittmann, *Modelling of GPCRs*,

DOI 10.1007/978-94-007-4596-4_8,

one significant problem: All experimentally detected GPCR-G protein interactions cannot be explained by only one model (Oldham et al. 2008) Thus, two hypotheses are suggested: First receptor dimers might play a role in interaction with G proteins or secondly, a sequential binding model is suggested (Herrmann et al. 2004, 2006).

Besides the experimental studies, a distinct number of modelling studies with regard to GPCR-G protein interactions are available (Fanelli et al. 1999; Greasley et al. 2001; Oliveira et al. 2003; Chou 2005; Raimondi et al. 2008; Strasser and Wittmann 2010b, in press). As already mentioned before, meanwhile, the molecular dynamic simulation of GPCRs in its natural surrounding is state of the art. However, it should be taken into account, that in case of simulating a ligand-receptor-complex in its active state, a pocket in the intracellular part of the GPCR is widely open. Thus, if the intracellular part of the GPCR is not in contact with the Gα subunit, but instead with water, this might lead to artefacts in simulations. Thus, the inclusion of the G protein, or at least a part of it, during active state simulations is recommended. A crystal structure of opsin, cocrystallized with the 11 last amino acids of the C-terminus of Gα is available (Scheerer et al. 2008). Thus, by homology modelling this crystallized system, inclusive the part of the C-terminus of Gα, can be adopted to the interesting system. However, the 11 last amino acids of the C-terminus of Gα represent only a very small part of the whole Gα and there are still important regions of interactions missing. As already mentioned in Chap. 2, in 2011, a crystal structure of the hβ_2R-G$\alpha\beta\gamma$-complex, which is shown in Fig. 2.7 was published for the first time (code: 3SN6, http://www.pdb.org) (Rasmussen et al. 2011). Despite the missing of a complete I3-loop and complete C-terminus of the hβ_2R in the crystal structure 3SN6 (http://www.pdb.org), this crystal gives a snapshot of one GPCR-G$\alpha\beta\gamma$-complex, and thus a more detailed insight onto the interaction between a GPCR and a G protein on molecular level. However, due to the hypothesis of sequential binding (Herrmann et al. 2004), as mentioned above, more different GPCR-G protein complexes should be taken into account.

The following part of Chap. 8.1 is mainly based on articles in literature (Strasser and Wittmann 2010b, in press; Copyright by Springer, with permission from Springer).

In general, two strategies for modelling a GPCR-G protein complex are possible: The most simple strategy would be the homology modelling of the complete GPCR-G protein complex based on the crystal structure 3SN6 (Rasmussen et al. 2011). Alternatively, the modeller can try to dock the Gα-subunit to the corresponding intracellular part of the receptor manually. But it is very hard to find an optimal complex and during this manual docking process a lot of clashes between GPCR and Gα subunit may occur. Furthermore by manual docking a large number of GPCR-G protein complexes can be received, and the modeller needs a criterion to decide which complex is the best one. Shortly, a manual docking is very unsystematically, and it is not recommended. Instead, a systematic search will lead in an easier way to better results. Thus, a procedure for systematic search should be introduced now: First, a homology model of the interesting GPCR in its active conformation should be generated, as already shown (see Chap. 3). Additionally, a homology model of the corresponding Gα subunit is needed. The homology model of the Gα subunit can

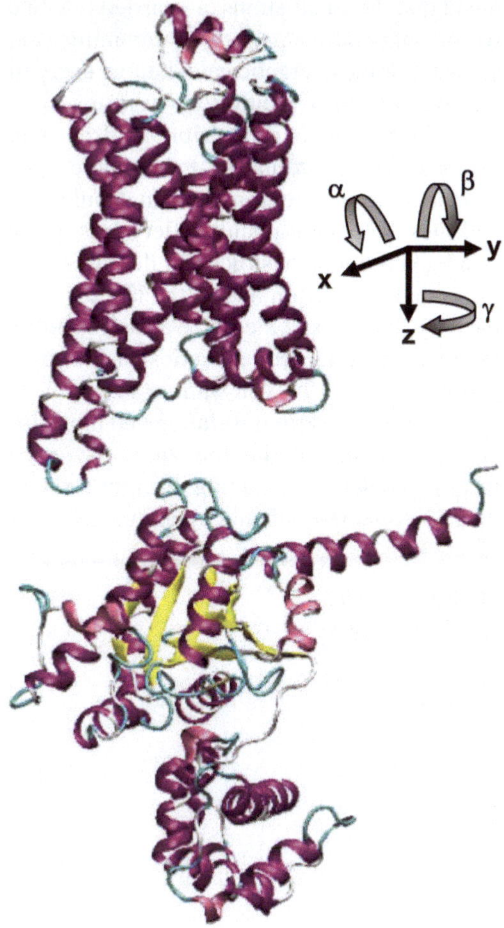

Fig. 8.1 Starting structure for the surface scan between a GPCR and the Gα subunit. (Copyright by Springer, with permission from Springer)

be generated in a similar manner, as described for the GPCRs. For the systematic search, a starting structure of the GPCR-G protein complex is needed. Optimally, the starting structure should be modelled in the following way:

1. The vertical axis of the receptor should be aligned in z direction of a coordinate system, thus the interaction surface of the GPCR with the Gα is found in a xy plain (Fig. 8.1).
2. Now the Gα has to be positioned in an optimal manner. Put it in a distinct distance below the intracellular part of the receptor in such way that the C-terminus of Gα points into direction of the open pocket of the receptor in the intracellular (Fig. 8.1). There should be no contact between the sites of the receptor and the Gα subunit.

There is no software for systematic scan of the potential energy surface available. Thus, the modeller has to establish the corresponding software by his own. We recommend that the calculations are carried out on a computer with LINUX. Furthermore we recommend using the programming language C in combination with C-shell scripts. For the energetic calculations every modelling software can be used in general, but GROMACS (http://www.gromacs.org) is the most fast and most flexible one.

A short schematic description of an appropriate source code is illustrated in Fig. 8.2. The coordinates of the whole system, i.e. GPCR and Gα-subunit must be read in. In the program code, it should be separated between sites of the receptor and sites of the Gα-subunit. Afterwards, you have to construct an architecture of six interlocking loop constructs. In the first three loops, the Gα-subunit is translated in x-, y- and z-direction and in the last three loops, the Gα-subunit is rotated around the x-, y- and z-axis. Please be aware, that the loops must be ordered in an interlocking manner. A subsequent series each after the other does not result in the desired systematic search. If you want to search on 10 points on each of the six dimensions (three dimensions for translation and three dimensions for rotation), you have to calculate 10^6 points. Within the 6th loop, you have to write out the coordinates of your resulting structure, including receptor and Gα-subunit. Be aware, that only the coordinates of the Gα-subunit were changed. Afterwards, you can call the GROMACS

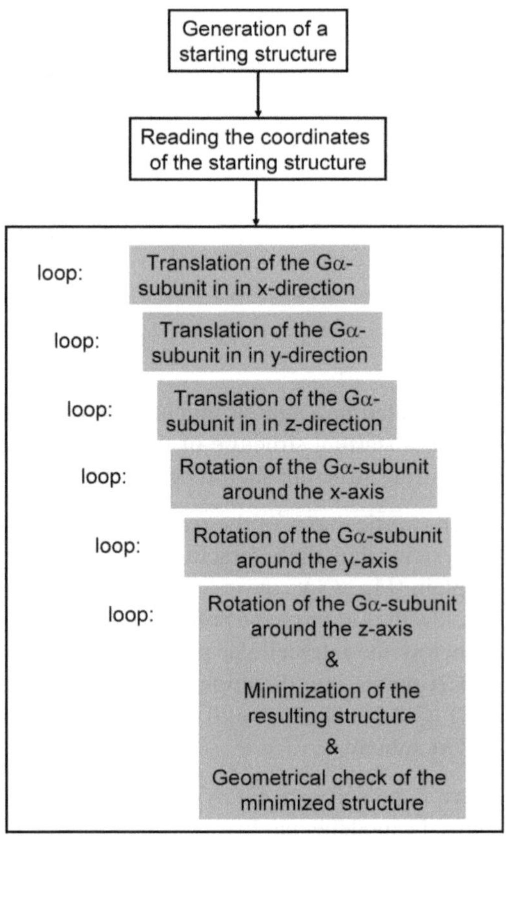

Fig. 8.2 Schematic presentation of a procedure to perform a systematic surface scan between a GPCR and a Gα-subunit

minimization within your C-code. After minimization, you can go back in your C-code, determine the potential energy of the minimized structure and save the potential energy in an appropriate data structure. Additionally, we strongly recommend to perform a geometrical check of each minimized structure. Especially, in the case, where the Gα-subunit is very close to the receptor, collisions, which were not cleaned by minimization, can occur. It may take some time, until the program code works quite fine, but a full automatization of all steps, due to the large number of structures is necessary and cannot be performed manually. Furthermore, we strongly recommend to split the program code into several modules. For example, one can establish a function for translation, a function for rotation and a function for calling a shell-script with the GROMACS routines for minimization.

As a result of the potential surface scan, one obtains the corresponding potential energy surfaces. A section of the potential surface with regard to translation on z-axis and rotation around z-axis is given in Fig. 8.3.

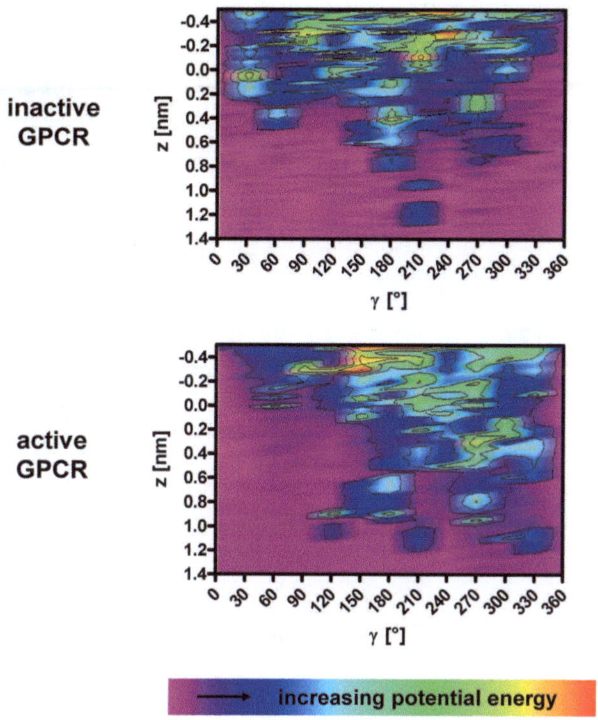

Fig. 8.3 Section of the potential surface, describing the interaction between a receptor and the Gα-subunit with regard to translation on z-axis and rotation around z-axis. (Copyright by Springer, with permission from Springer; modified)

Based on the potential energy surface, shown in Fig. 8.3 a model for a hβ_2R-Gα_s-complex was suggested (Strasser and Wittmann 2010b). Because the crystal structure of a hβ_2R-Gαβγ-complex was published recently (Rasmussen et al. 2011), the prediction was compared with the experimentally determined structure (Strasser and Wittmann, in press). The comparison between predicted (model I) and experimental results revealed an rmsd of about 8.4 Å. Therefore, the potential energy surface scan was extended in order to find a structure with smallest rmsd compared to the crystal structure (Fig. 8.4).

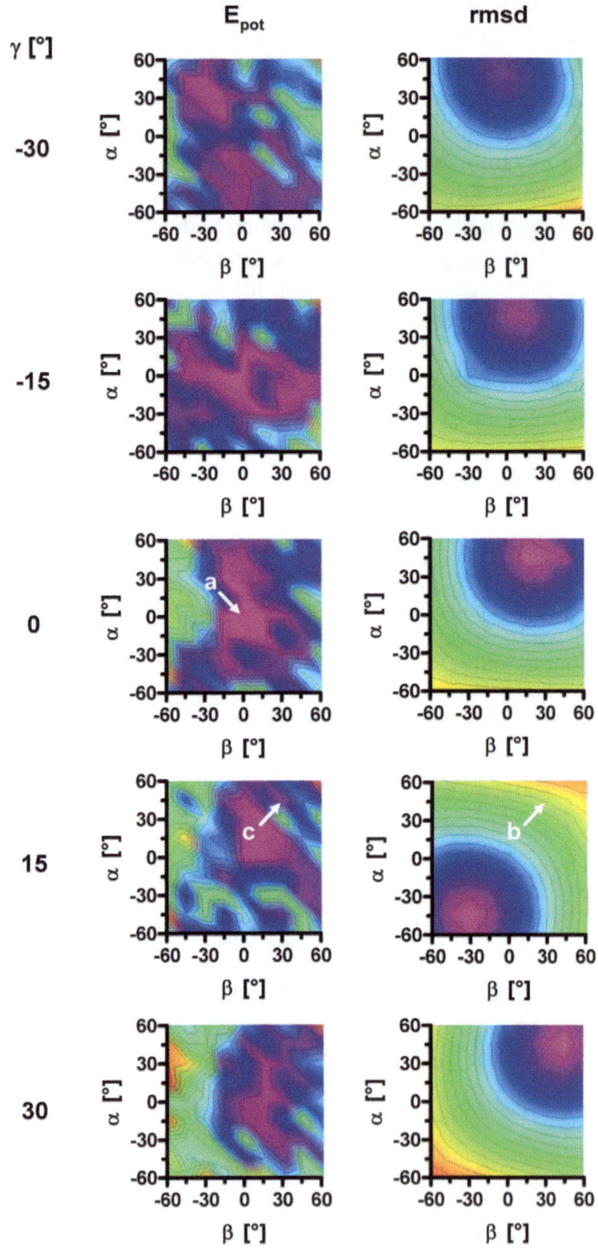

Fig. 8.4 Potential energy (E_{pot}) and *rmsd* surfaces for the systematic search in the range α, $\beta = -60°$ to $60°$ and $\gamma = -30°$ to $30°$. *arrow a*: model I, representing a gobal minimum on the potential energy surface as predicted (Strasser and Wittmann 2010b); *arrow b*: minimum *rmsd* between the calculated model I and the corresponding parts of the crystal structure; *arrow c*: local minimum on the potential energy surface, representing the smallest *rmsd* with regard to the crystal structure. (Copyright by Springer, with permission from Springer)

A local minimum on the potential energy surface (model Ia) with an rmsd between model and crystal structure of about 3.3 Å was identified (Fig. 8.4). An alignment of the predicted models I and Ia with the crystal structure is shown in Fig. 8.5.

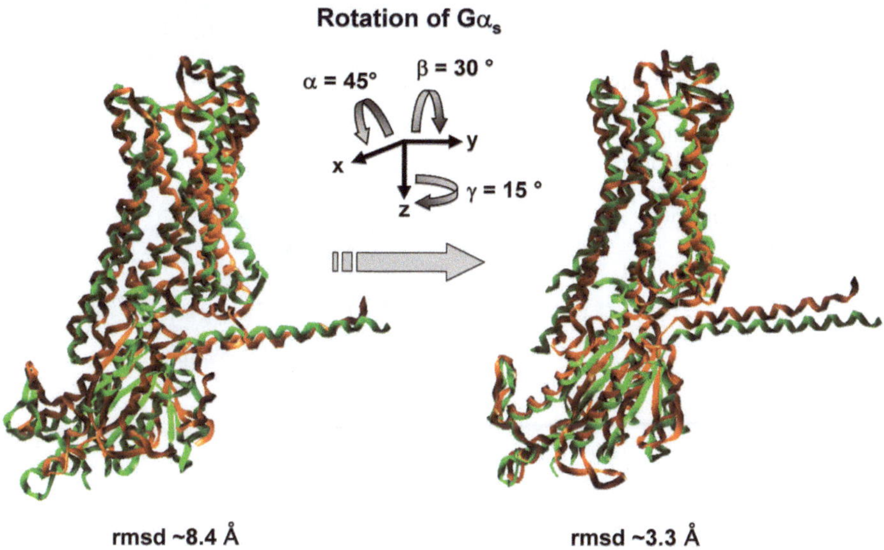

Fig. 8.5 Alignment of model I (*left*, *orange*) or model Ia (*right*, *orange*) with the corresponding parts of the crystal structure 3SN6 (*green*). (Copyright by Springer, with permission from Springer)

Taking into account that the crystal structure is artificial due to the cocrystallization of the Gs-binding nanobody (Nb35), and the T4 lysozyme (Fig. 2.7), the predicted model I is in very good accordance to the experimental structure. As already stated, all experimental results concerning GPCR-G protein interaction, including mutagenesis and pharmacological studies cannot be explained by only one interaction model (Oldham and Hamm 2008). Therefore, the hypothesis of sequential binding is discussed in literature (Herrmann et al. 2004). This hypothesis may be supported by the modelling results, because a sequential binding pathway, connecting model I and Ia, was determined on the potential energy surface (Fig. 8.6).

As crystal structures are snapshots of distinct conformations in the solid state, molecular modelling studies afford insight into distinct amino acid interactions between the receptor and Gα not only for minima, but also for intermediate states, which cannot be obtained via crystal structures. Thus, molecular modelling studies may allow deeper insights onto binding mechanism of a Gα to a GPCR.

Fig. 8.6 Potential energy surfaces of the predicted hβ_2R-Gα_s-complex and the minimum energy pathway. **a** Minimum energy pathway connecting model Ia (*point 1*) and model I (*point 8*). **b** Schematic presentation of the minimum energy pathway connecting model Ia (*point 1*) and model I (*point 8*) along with the corresponding angles α, β and γ. (Copyright by Springer, with permission from Springer)

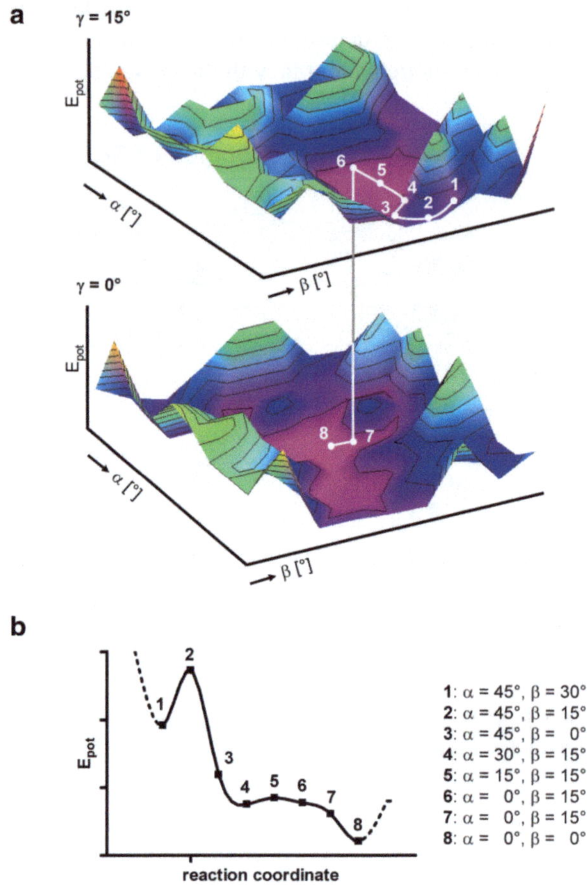

8.2 Process of Ligand Binding from the Extracellular Side into the Binding Pocket of a GPCR

It is widely accepted, that ligands bind deeply between the transmembrane domains of for example, biogenic amine receptors. On the one hand, this is suggested by crystal structures, cocrystallized with a ligand in the binding pocket. On the other hand, this fact is supported by site-directed mutagenesis studies. However there is only very few knowledge about the ligand-recognition on receptor surface and the guiding of the ligand into the binding pocket of a GPCR. With experimental studies, the whole binding process on molecular level can not be studied in detail. In general, molecular modelling studies are able to give these insights on molecular level. Since the ligand binding is of dynamic nature, one would think about using molecular dynamic simulations to study the binding process of a ligand. However, until now, no study, observing the complete binding process of a ligand from the extracellular side into the binding pocket of a receptor was published. The reason for this lack is the large computing time, due to the time scale of ligand binding process. Some studies used the technique of steered molecular dynamics (Isralewitz et al. 1997; Kosztin et al. 1999). In order to get more insight onto the ligand binding process on molecular level, the algorithm "*LigPath*" is described in literature as an alternative method (Strasser and Wittmann 2007a, b, 2010).

The following part of Chap. 8.2 is mainly based on articles in literature (Strasser and Wittmann 2007b, 2010; Copyright by Springer, with permission from Springer).

Antagonists or inverse agonists are suggested to stabilize the inactive conformation of the receptor, whereas (partial) agonists stabilize the active conformation of a receptor. Thus, it can be suggested, that during the binding process of the (partial) agonist, the receptor has to change its conformation. In general, for the binding process of a (partial) agonist, several pathways, as illustrated in Fig. 8.7, have to be taken into account.

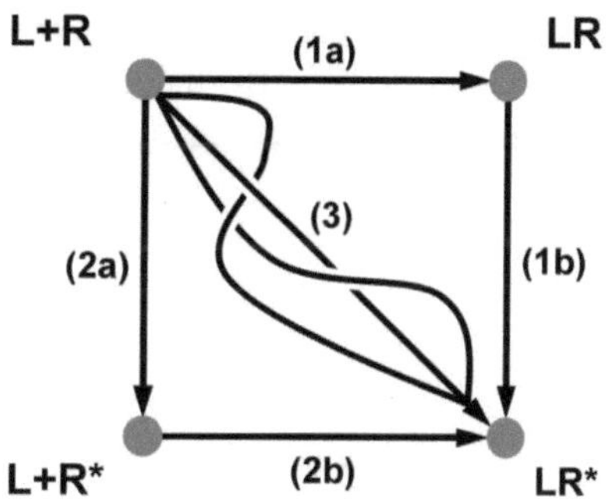

Fig. 8.7 Scheme for different ligand-binding and receptor-activation pathways. (Copyright by Springer, with permission from Springer)

The binding of an antagonist or inverse agonist (L) to the receptor (R), is illustrated by pathway (1a) of Fig. 8.7. After binding, the inactive ligand-receptor-complex (LR) is established. For binding of a (partial) agonist, three different pathways have to be taken into account:

- Pathway 1a & 1b
  The (partial) agonist (L) binds to the inactive state of the receptor (R) and establishes the inactive ligand-receptor-complex (LR). Subsequently, the inactive ligand-receptor-complex (LR) changes its conformation into the active ligand-receptor-complex (LR*)
- Pathway 2a & 2b
  The inactive receptor (R) changes its conformation into the active state (R*) without binding of a ligand. This phenomena is called "constitutive activity". Subsequently, the (partial) agonist (L) binds to the already activated receptor (R*) and the active ligand-receptor-complex (LR*) is established
- Pathways 3
  The (partial) agonist (L) starts to bind to the inactive conformation of the receptor (R), but during the ligand-binding process, the receptor gets activated and the active ligand-receptor-complex (LR*) is established

As pointed out, the receptor activation can take place during different states of ligand-binding. In order to get more detailed insights into ligand-induced receptor-activation, *LigPath*-calculations can be performed.

For such a calculation, a starting and a destination structure is needed. The starting structure may be defined by the inactive receptor embedded in its natural surrounding, like lipid bilayer, intra- and extracellular water, and the ligand somewhere in the aqueous phase of the extracellular side. In contrast, the destination structure may be defined by the active ligand-receptor-complex, embedded in its natural surrounding. Both, starting and destination structure can be obtained by homology modelling. Subsequently, both models should be embedded into the appropriate surrounding and molecular dynamic simulations (see Chap. 6) should be performed, in order to obtain a well equilibrated starting and destination structure.

The aim of the *LigPath*-calculation is to get deeper insight into the activation process during binding of a (partial) agonist. As pointed out in Fig. 8.7, several pathways (1a and 1b, 2a and 2b and 3) have to be considered. Consequently, a complete systematic scan of the potential energy surface, as shown schematically in Fig. 8.8 has to be performed.

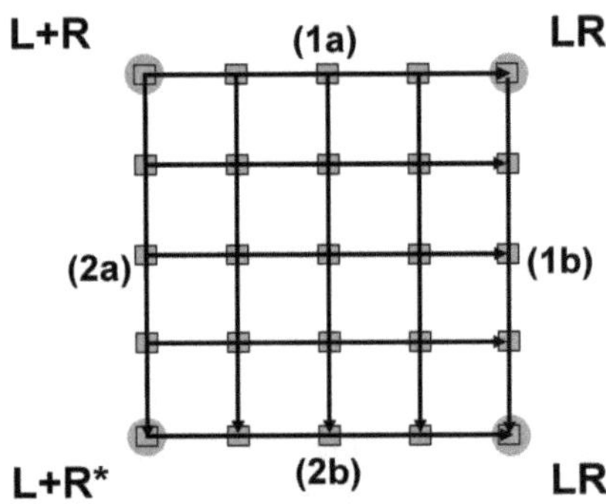

Fig. 8.8 Schematic presentation of a systematic surface scan; the long *black arrows* indicate the direction for the pathway calculation, whereas the *white* and *black* boxes represent schematically the lattice points used for surface calculation. (Copyright by Springer, with permission from Springer)

To reduce the computation time, the *LigPath*-algorithm can be used alternatively to such a systematic scan. Specific for the *LigPath*-algorithm is the generation-child-scheme, as illustrated in Fig. 8.9.

Therein, in each cycle, also named "generation", of the calculation, three different groups of child structures are calculated. In the first group (Fig. 8.9, (I)), only the ligand atoms are guided differentially in direction of their destination position. The guiding of the ligand atoms is combined with a Monte-Carlo-like procedure, so that the guiding also has a random character. In the second group (Fig. 8.9, (II)), only the atoms of the receptor are guided differentially in direction of their destination position. As for the ligand atoms, the guiding of the receptor atoms is combined with a Monte-Carlo-like procedure, thus, the guiding has a random character again. In the third group (Fig. 8.9, (III)), ligand as well as receptor atoms are differentially

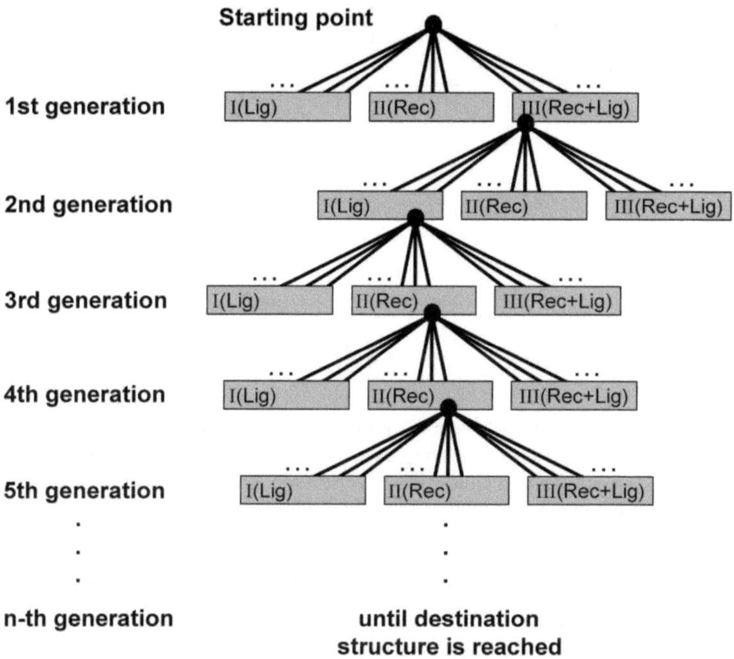

Fig. 8.9 Generation-child scheme of the *LigPath*-calculation. Based on the starting structure, new, minimized child-structures are generated by the *LigPath*-algorithm. The children are divided into three groups *I*, *II* and *III*. Each group contains *n* children. The best child of each generation is used as starting structure for the next generation. The generation-child-cycle is continued, until the destination structure is obtained. (Copyright by Springer, with permission from Springer)

guided in direction of their destination position. Again, due to the combination with a Monte-Carlo-like procedure, the guiding has a random character.

The division of each generation into three different child-groups is very important in order to obtain a non-restrained ligand-binding and receptor-activation pathway. Out of each generation, the "best" child is used as starting structure for the next generation. The "best" child of each generation is selected by a combined criterion with regard to "rmsd" and "potential energy". On the one hand, the potential energy of the best child should be as small as possible, on the other hand, the rmsd between the actual structure and the destination structure should be as small as possible, too. This criterion guarantees on the one and, that only structures with low potential energies are chosen, but on the other hand, the structure is guided, because of the different child-groups, without restraints into direction of the destination structure, as pointed out in Fig. 8.10. Based on the combined "energy-rmsd"-criterion, mentioned above, the best structure can be related with movement of ligand only, with movement of receptor only, or with movement of ligand and receptor. Thus, the pathway, starting from the free ligand and free inactive receptor (L + R) forward to the active ligand-recetor-complex (LR*) is not restrained. This fact is very important, in order to get knowledge, at which stage of ligand penetration, the receptor gets activated.

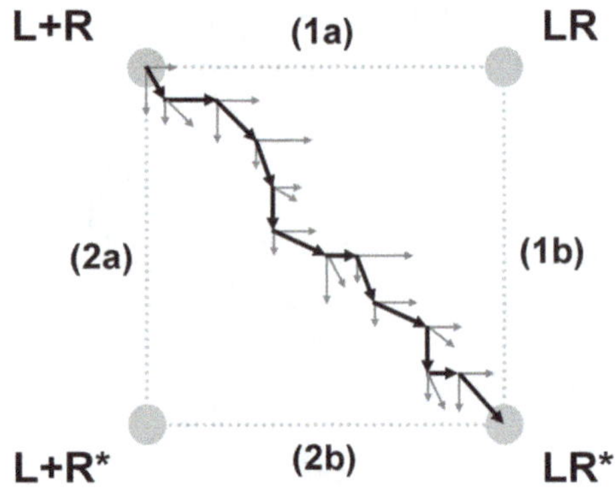

Fig. 8.10 Schematic presentation of a non-restrained *LigPath*-calculation: the three *arrows*, beginning for each generation in the same origin represent the three groups of children in each generation; the *black arrows* represent the best child of each generation; the final point of each *black arrow* is the starting point for the next generation. (Copyright by Springer, with permission from Springer)

For the binding of a partial agonist to a biogenic amine receptor, the binding pathway was calculated with the *LigPath*-algorithm (Fig. 8.11, dotted line). Besides that, a systematic potential energy surface scan, presented in Fig. 8.11, was performed. The potential surface scan reveals a minimum energy path that is in good accordance to the minimum energy pathway, calculated by *LigPath* (Fig. 8.11, dotted line). Thus, the *LigPath*-algorithm can be used as alternative to a systematic energy surface scan. This is advantageous with regard to a decreased computation time. However, the systematic surface scan gives more detailed insights onto the potential energy

Fig. 8.11 Potential energy surface for penetration of a partial agonist into the binding-pocket of a biogenic amine receptor. (Copyright by Springer, with permission from Springer; modified)

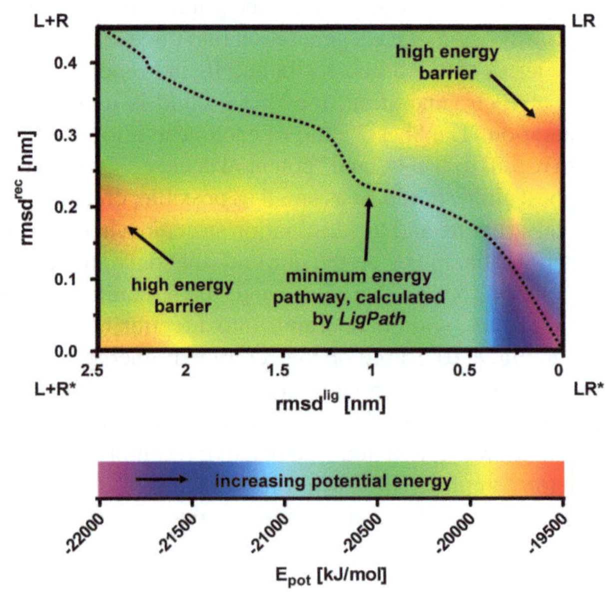

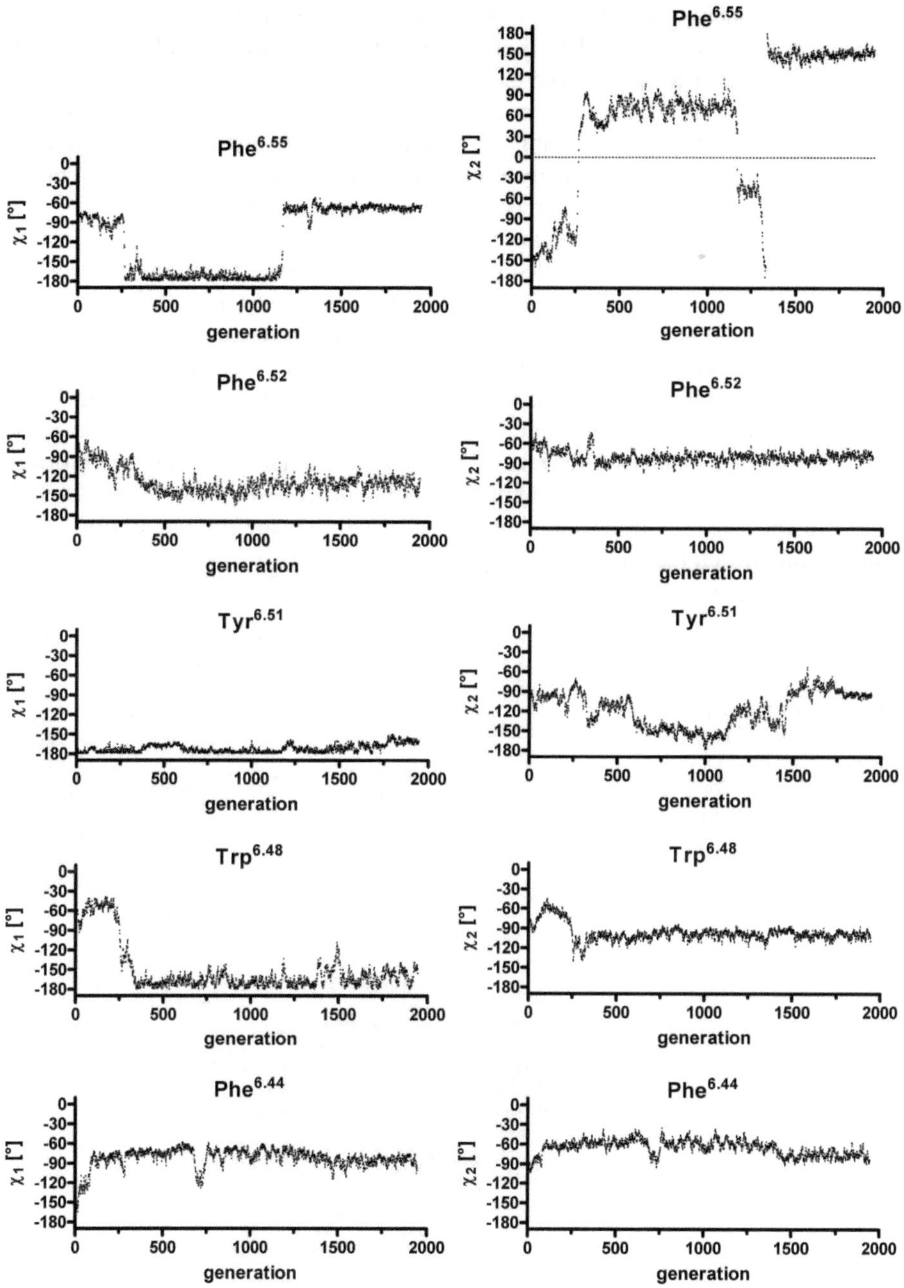

Fig. 8.12 Changes in dihedral angles of distinct amino acid side chains during binding of a (partial) agonist. (Copyright by Springer, with permission from Springer)

surface: The inactive receptor (R) is separated from the active receptor (R*) by a high potential energy barrier. This is also true for the inactive (LR) and active ligand-receptor-complex (LR*). However, the high energy barrier shows a small pass, which can be passed during the binding process of the ligand. Furthermore, this modelling data suggest, that the ligand binding, accompanied by receptor activation is, at least for this example, the energetically preferred pathway. Thus, during the binding process of a (partial) agonist, the receptor gets more and more activated.

As described above, the *LigPath*-calculations allow getting insight onto potential energy surface, but also detailed insights onto processes on structural level can be obtained. Largest changes, for example, were described for $Phe^{6.55}$ and $Trp^{6.48}$ (Fig. 8.12). The conformational changes of the dihedral angles $\chi 1$ and $\chi 2$ of the aromatic amino acid side chains ($Phe^{6.55}$, $Phe^{6.52}$ and $Tyr^{6.51}$) establish an aromatic channel for the phenylmoieties of the ligand. Due to the binding process of the ligand (Fig. 8.13, (1)), the ligand gets in close contact to $Phe^{6.55}$. This induces a cascade of conformational changes (Fig. 8.13, (2)–(5)). $Phe^{6.55}$ undergoes a large conformational change in the early beginning of the binding process (Fig. 8.12). The dihedral angles of the $Phe^{6.55}$ side chain change significantly during the process of ligand binding and switch back into their original conformation, after the ligand is bound to the binding pocket.

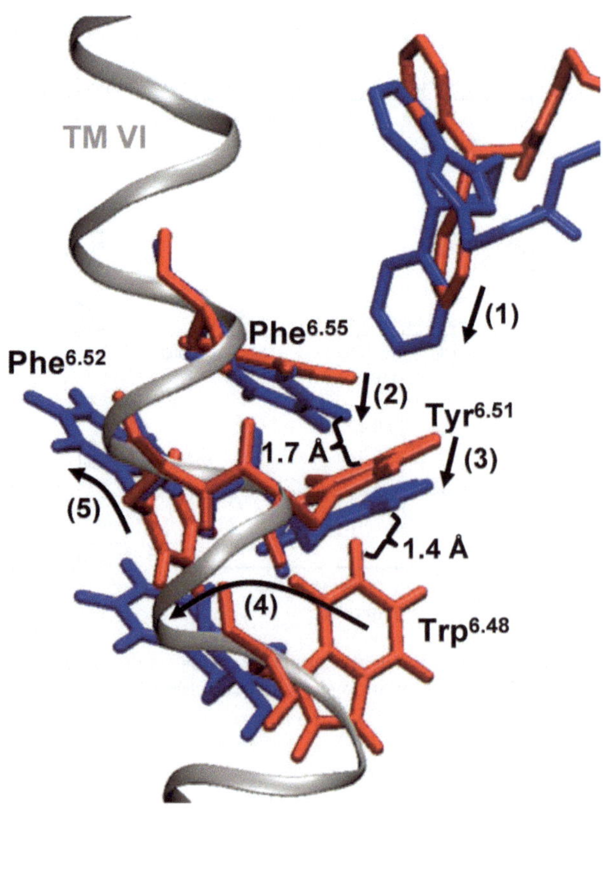

Fig. 8.13 Sequence of conformational changes of distinct amino acid side chains during binding of a (partial) agonist. (Copyright by Springer, with permission from Springer)

During the whole phase of ligand penetration, further changes in dihedral angles of Phe$^{6.52}$, Tyr$^{6.51}$, Trp$^{6.48}$ and Phe$^{6.44}$ could be observed (Fig. 8.12).

However, changes during ligand penetration and receptor activation do not take place only in the upper part of the transmembrane domains, but also in the intracellular part. Some representative changes of amino acid conformations within the H_1 receptor during activation are shown in Figs. 8.14 and 8.15.

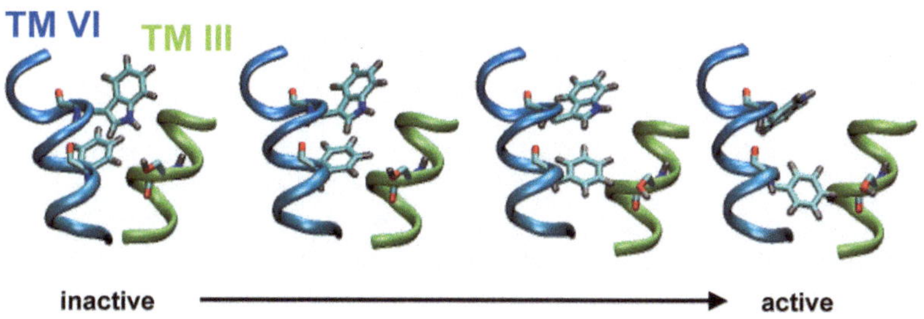

Fig. 8.14 Snapshots of Trp$^{6.48}$, Phe$^{6.44}$ and Ser$^{3.39}$ during the activation process of the gpH_1R. (Copyright by Springer, with permission from Springer)

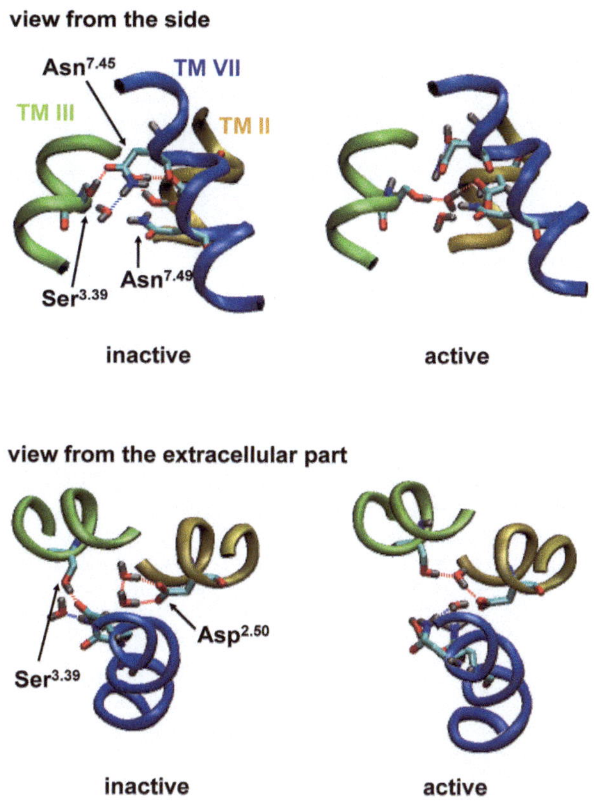

Fig. 8.15 Ser$^{3.39}$-switch and three conserved internal water molecules between *TM II*, *TM III* and *TM VII* in the intracellular part of the transmembrane helix bundles of the gpH_1R. (Copyright by Springer, with permission from Springer)

In general, the *LigPath*-algorithm allows studying structural and energetical changes during ligand binding and receptor activation on molecular level in detail. Thus, it is really worth to implement such an algorithm. However, one should taken into account, that therefore it is necessary to master a programming language, like C. Additionally, one should be able to perform some geometrical calculations, like rotation around axis, rotation around a bond, translation and similar.

Chapter 9
Force Fields

Taking into account the computational difficulties as mentioned in the introductionary part, when calculating the potential energy of a system, the use of the so-called force fields enables us to get relevant structural and energetic information. In this chapter, the most important facts, concerning force fields will be presented. For more detailed information, the reader is referred to appropriate literature (Halgren et al. 1996; Jorgensen et al. 1996; MacKerell et al. 1998; Jensen 1999; Duan et al. 2003; Mackerell 2004; Oostenbrink et al. 2004; van der Spoel et al. 2005; Kukol 2010).

9.1 The Force Field Energy

The potential energy of a system, also called force field energy is given in Eq. 9.1.

$$E_{FF} = E_{bond} + E_{angle} + E_{tors} + E_{vdW} + E_{el} + E_{cross}. \tag{9.1}$$

Therein, E_{bond} describes the energy function for stretching a bond between to atoms, E_{angle} is the energy function describing the bending of an angle between three atoms. The torsional energy is given by E_{tors} and describes the energy for rotation around a bond. E_{vdW} and E_{el} represent the non-bonded van der Waals and electrostatic interactions. The coupling between the stretching, bending and torsional energy is described by the cross term E_{cross}.

9.1.1 The Stretching Energy

The stretching energy E_{str} represents the energy function for stretching a bond between two atoms (Fig. 9.1).

This energy can be described by Eq. 9.2:

$$E_{bond} = k^{AB}\left(r^{AB} - r_0^{AB}\right)^2 \tag{9.2}$$

A. Strasser, H.-J. Wittmann, *Modelling of GPCRs*,
DOI 10.1007/978-94-007-4596-4_9, © Springer Science+Business Media Dordrecht 2013

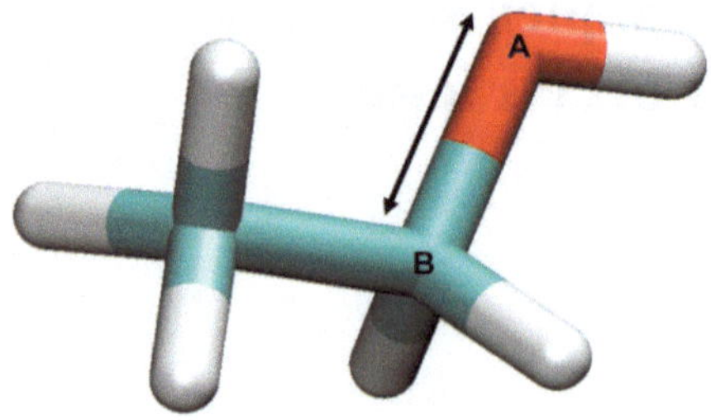

Fig. 9.1 Definition of a bond

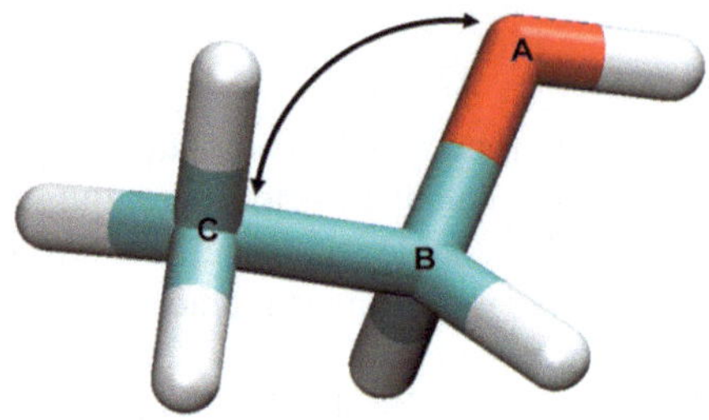

Fig. 9.2 Definition of the angle

r_0^{AB} reference distance between both atoms A and B
r^{AB} actual distance between the atoms A and B
k^{AB} force constant for the bond between A and B.

Besides this simple equation, some other equations to describe the stretching energy are used in literature. Because of computational efficacy, in the GROMOS-96 force field (van Gunsteren et al. 1996) the stretching energy is described by the following Eq. 9.3:

$$E_{bond} = \frac{1}{4} k^{AB} \left(\left(r^{AB} \right)^2 - \left(r_0^{AB} \right)^2 \right)^2. \tag{9.3}$$

9.1.2 *The Bending Energy*

The bending energy E_{angle} describes the bending of an angle between the three atoms A, B and C with a bond between A and B and between B and C (Fig. 9.2).

The bending energy can be represented by the following harmonic approximation 9.4:

$$E_{angle} = k^{ABC} \left(\alpha^{ABC} - \alpha_0^{ABC} \right)^2 \tag{9.4}$$

α_0^{ABC} reference angle between the atoms A, B and C
α^{ABC} actual angle between the atoms A, B and C
k^{ABC} force constant for the angle between A, B and C.

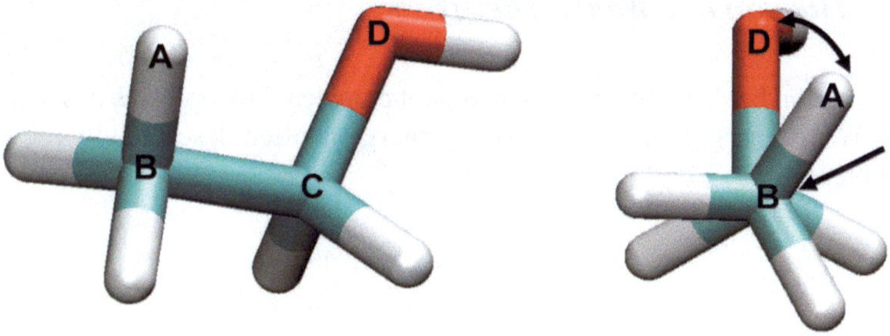

Fig. 9.3 Definition of the torsional angle

This simple harmonic approximation is sufficient enough for most problems. For more details, please have a look at the appropriate literature (Jensen 1999; van der Spoel et al. 2005). In the GROMOS-96 force field (van Gunsteren et al. 1996; van der Spoel et al. 2005), a simplified Eq. 9.5:

$$E_{angle} = \frac{1}{2} k^{ABC} \left(\cos\left(\alpha^{ABC}\right) - \cos\left(\alpha_0^{ABC}\right)\right)^2 \tag{9.5}$$

is used.

9.1.3 The Torsional Energy

The torsional energy, for rotation around the bond B-C within a four atoms A, B, C and D, connected by bonds between A and B, B and C, C and D (Fig. 9.3), is described by E_{tors}.

The corresponding energy term may be described by the Eq. 9.6:

$$E_{tors}(\omega) = \sum_n V_n \cos(n\omega) \tag{9.6}$$

ω dihedral (torsional) angle
n multiplicity
V_n barrier of rotation around the bond B-C.

The term n = 1 describes a rotation with 360° periodicity, the term n = 2 describes a rotation with 180° periodicity and the term n = 3 describes a rotation with 120° periodicity.

In GROMACS, the proper dihedral angles are defined according to the IUPAC/IUB convention (van der Spoel et al. 2005). Therein, ω is the angle between the plane ABC and the plane BCD. Zero corresponds to the cis configuration, this means, A and D are on the same side (van der Spoel et al. 2005). In GROMACS, the following Eq. 9.7 is used:

$$E_{tors}(\omega) = k\left(1 + \cos\left(n\omega - \omega_0\right)\right). \tag{9.7}$$

9.1.4 The van der Waals Energy

The interaction between atoms which are not connected by bonds is described by the van der Waals energy. The van der Waals energy is often described by the Lennard-Jones potential E_{LJ} Eq. 9.8:

$$E_{LJ} = \frac{C_{AB}^{(12)}}{r^{12}} - \frac{C_{AB}^{(6)}}{r^{6}} \tag{9.8}$$

$C_{AB}^{(12)}$, $C_{AB}^{(6)}$ parameters for the interaction between two atoms A and B
r actual distance between two atoms A and B.

Instead of the above equation, often a more prominent Eq. 9.9 is used

$$E_{LJ} = 4\varepsilon_{AB}\left[\left(\frac{\sigma_{AB}}{r}\right)^{12} - \left(\frac{\sigma_{AB}}{r}\right)^{6}\right] \tag{9.9}$$

ε_{AB} parameter for interaction between two atoms A and B
σ_{AB} parameter for interaction between two atoms A and B.

9.1.5 The Electrostatic Energy

The electrostatic interaction between two atoms, which are not bonded is represented by the following Eq. 9.10:

$$E_{el} = \frac{q_A q_B}{4\pi\varepsilon_0\varepsilon} \cdot \frac{1}{r} \tag{9.10}$$

q_A,q_B partial charges on the atoms A and B
r distance between the atoms A and B
ε_0 vacuum permittivity
ε dielectric constant of a medium.

9.2 The All-atom-concept and Site-concept

In general, there are two opposite concepts with regard to force-field-parameterization. On the one hand, there is the so-called all-atom model, e.g. ethanol (Fig. 9.4). As already indicated by the name, all atoms of the molecule are included into the calculation, and therefore, parameterized. On the other hand, there is the so-called site model (Fig. 9.4). Therein, a small group of atoms, in general connected via bonds, are summarized within one group, the so-called "site". It is very important, that the combination of several atoms into one site is ingenious. Thus, in general methyl moieties (CH_3), methylene moieties (CH_2) and aromatic and aliphatic CH

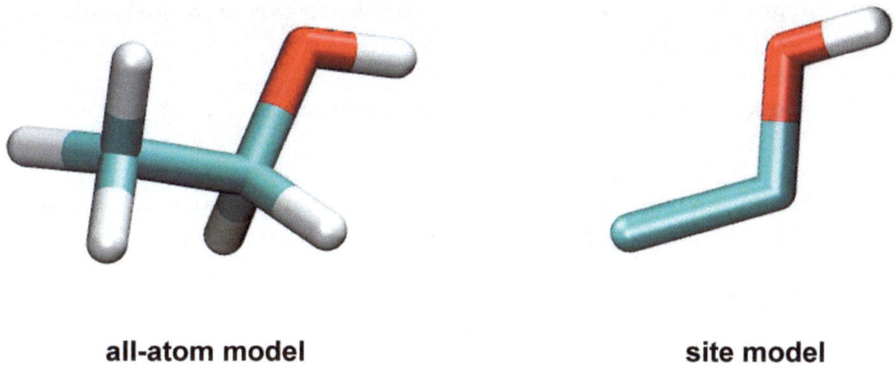

Fig. 9.4 All-atom model versus site model

moieties are summarized to one site. However, in some force-fields, aromatic CH moieties are not handled as one site.

Due to the dipole character, it would not be very useful, to combine the hydrogen and oxygen of a hydroxy moiety (OH) to one site. A combination into one site would not allow describing intermolecular interaction, like hydrogen bonds.

9.3 The Force Field Parameters

In literature, a number of force fields, which differ in the values for the non-variable parameters of the potential energy terms, mentioned in the foregoing section, are described. The most prominent for example are the following: TRIPOS (Clark et al. 1989), AMBER (Cornell et al. 1995; Duan et al. 2003), OPLS (Jorgensen et al. 1996), CHARMM (MacKerell et al. 1998) and GROMOS (Oostenbrink et al. 2004). Some force fields were developed for specific classes of molecules. For example, the EAS force field was developed for alkanes, whereas other force fields, like MM2 can be used in general. Other force fields were especially developed for proteins, nucleic acids and carbohydrates.

For performing any force-field based calculations, like minimization or molecular dynamic simulations, one needs the corresponding force field parameters. GROMACS for example, contains the command `pdb2gmx`. This command allows, if a correct pdb-file of a protein is available, to generate the so-called topology-file, which contains all information about the force-field parameters, necessary for simulation. Unfortunately, `pdb2gmx` can be used only for proteins. To obtain topology files for small compounds, the PRODRG-server (http://davapc1.bioch.dundee.ac.uk/prodrg/), as described in Chap. 4 can be used.

```
;
;
;       This file was generated by PRODRG version
        AA081006.0504
```

```
;      PRODRG written/copyrighted by Daan van Aalten
;      and Alexander Schuettelkopf
;
;      Questions/comments to dava@davapc1.bioch.dundee.
;      ac.uk
;
;      When using this software in a publication, cite:
;      A.W. Schuettelkopf and D.M.F. van Aalten (2004).
;       (2004).
;      PRODRG - a tool for high-throughput crystallography
;      of protein-ligand complexes.
;      Acta Crystallogr. D60, 1355--1363.
;
;
[ moleculetype ]
; Name nrexcl
  O2      3
[ atoms ]
;   nr   type   resnr resid  atom  cgnr   charge     mass
     1    CH3       1  LIG     C4     1    0.074  15.0350
     2    CH2       1  LIG     C1     1    0.091  14.0270
     3     OA       1  LIG     O2     1   -0.202  15.9994
     4      H       1  LIG     H2     1    0.037   1.0080

[ bonds ]
; ai aj fu    c0, c1, ...
   2  1  2    0.153  7150000.0  0.153  7150000.0 ; C1  C4
   2  3  2    0.143  8180000.0  0.143  8180000.0 ; C1  O2
   3  4  2    0.100 15700000.0  0.100 15700000.0 ; O2  H2

[ pairs ]
; ai aj fu    c0, c1, ...
   1  4  1                                       ;  C4 H2

[ angles ]
; ai aj ak fu  c0, c1, ...
   1  2  3  2   109.5  520.0  109.5  520.0 ;   C4  C1 O2
   2  3  4  2   109.5  450.0  109.5  450.0 ;   C1  O2 H2

[ dihedrals ]
; ai aj ak al fu  c0, c1, m, ...
   1  2  3  4  1  0.0 1.3 3 0.0 1.3 3 ; dih C4 C1  O2 H2
```

In the section `[ bonds ]`, the parameters for the stretching energy between two bonded atoms are defined. Let's look onto the first parameter line of this section:

```
   2  1  2  0.153  7150000.0  0.153  7150000.0 ;  C1 C4
```

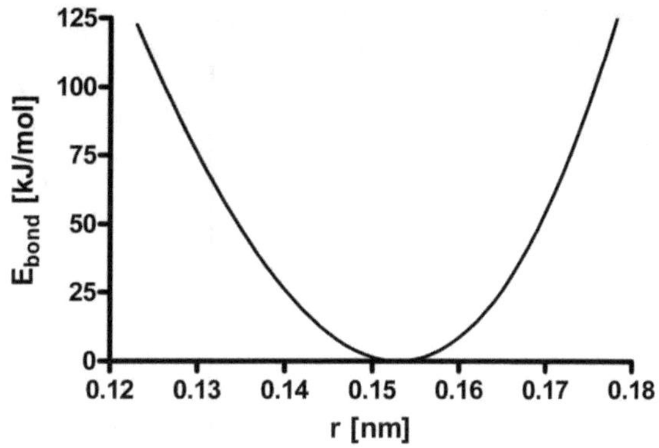

Fig. 9.5 E_{bond} in dependence of the distance between two atoms (example: bond between C1 and C4 of ethanol, see above)

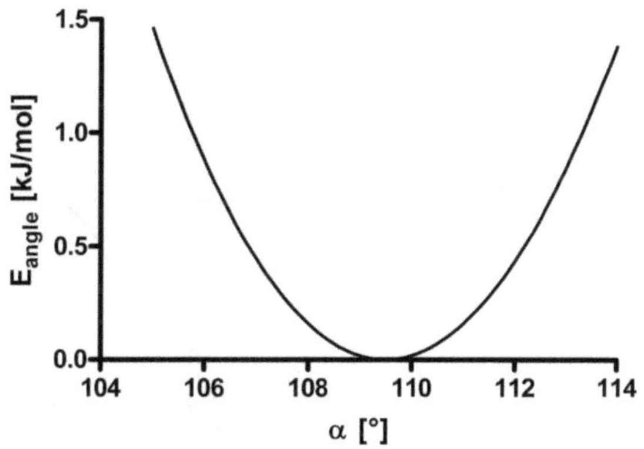

Fig. 9.6 E_{angle} in dependence of the angle between three atoms (example: angle between C4, C1 and O2 of ethanol, see above)

Within this line, the stretching energy between the second site (C1 = CH2) and the first site (C4 = CH3) is defined. There, r_o^{AB} is related with the value of 0.153 nm, whereas the force constant k^{AB} is related with the value of 7150000.0 kJ/(mol nm^4). The related dependence of E_{bond} in dependence of the distance is given in Fig. 9.5.

In the section `[ angles ]`, the parameters for the bending energy between three bonded atoms are defined. Let's look onto the first parameter line of this section:

```
1  2  3  2  109.5  520.0  109.5  520.0 ;  C4  C1  O2
```

Within this line, the bending energy between the first (C4 = CH3), second (C1 = CH2) and the third site (O2 = OA) is defined. There, $\alpha_o^{ABC} = 109.5°$ and $k^{ABC} = 520.0$ kJ/mol. The corresponding course of the bending energy is shown in Fig. 9.6.

In the section `[ dihedrals ]`, the parameters for the torsional energy between four atoms are defined. Let's look onto the first parameter line of this section:

```
1 2 3 4 1 0.0 1.3 3 0.0 1.3 3 ; dih C4 C1 O2 H2
```

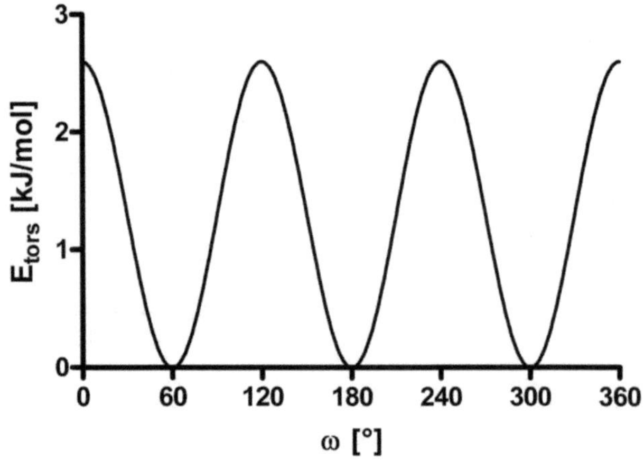

Fig. 9.7 E_{tors} in dependence of the torsion angle (example: dihedral angle between C4, C1, O2 and H2 of ethanol, see above)

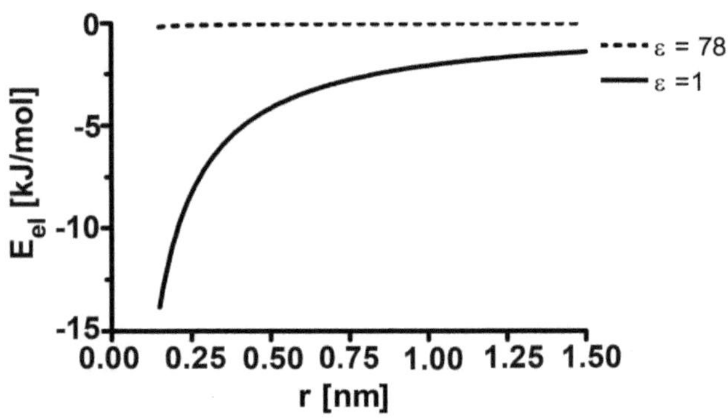

Fig. 9.8 E_{el} in dependence of the distance r between two atoms (example: coulomb interaction between C4 and O2 of ethanol, see above)

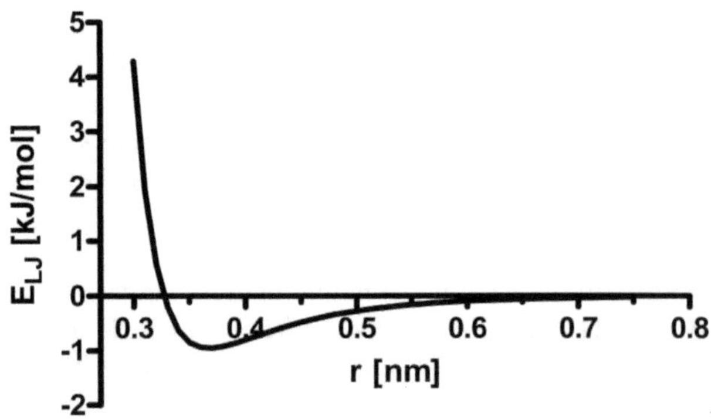

Fig. 9.9 E_{LJ} in dependence of the distance r between two atoms (example: Lennard-Jones interaction between C4 and O2 of ethanol, see above)

Here the torsional energy between site 1 (C4 = CH3), site 2 (C1 = CH2), site 3 (O2 = OA) and site 4 (H2 = H) of ethanol is described. For the presented example ω_o is 0°, the force constant k is 1.3 kJ/mol and the multiplicity n is 3. The corresponding course of the torsional energy is shown in Fig. 9.7.

There is no separate section with regard to partial charges of the sites. The partial charges in the topology files of GROMACS are found in the section [atoms] in the 7th column. Let us look for example onto the partial charge of the CH3- and OA-site: For the CH3-site, the partial charge is 0.074 e_0 (e_0 denotes the elementary charge), whereas the partial charge for the OA-site is -0.202 e_0, as suggested in the corresponding topology file (see above). The electrostatic interaction between the CH3- and OA-site in dependence of distance r for vacuum ($\varepsilon = 1$) and aqueous phase ($\varepsilon = 78$) is presented in Fig. 9.8.

The Lennard-Jones or van der Waals parameter are not found within the topology file, generated by the PRODRG server. Instead the are found in the corresponding parameter file. For the `ffG53a6` force field (Oostenbrink et al. 2004) for example, the Lennard-Jones parameters are found in the file `ffG53a6nb.itp`, which may be located in the directory `gromacs/share/gromacs/top` in your GROMACS installation directory. For the interaction between a CH3- and OA-site, a value of 0.004663258 kJ/(mol nm^6) is mentioned for the $C^{(6)}$- and a value of $5.6782 \cdot 10^{-6}$ kJ/(mol nm^{12}) for the $C^{(12)}$-parameter. The Lennard-Jones interaction between the CH3- and OA-site in dependence of distance r is presented in Fig. 9.9.

Chapter 10
Thermodynamics of Ligand-Receptor Interaction

10.1 Motivation

The development of a new drug nowadays spents a lot of time and money starting with the synthesis of the molecule and ending up in the testing process. Very often promising compounds built up by, for example screening methods, fail to exhibit the desirable properties. A better understanding of the interaction process for the ligand-receptor system may circumvent these problems and may allow to design drug purposeful in future. An extensive discussion of thermodynamical concepts, presented in this chapter is given in literature (Kondepudie et al. 1998; Silberg et al. 2005; Klotz et al. 2008).

10.2 Ligand-Receptor Model

A lot of work has done during the last decades to investigate the interaction between a ligand and its receptor, resulting in the proposal of forming a ligand-receptor complex as a fundamental step determining the effect of a drug (Kenakin 1997). In its simplest form, this process will be described as an equilibrium between the ligand located in the extracellular "solution", the receptor embedded in the cell membrane and the complex where the ligand occupies a binding pocket inside the receptor:

$$\mathrm{L} + \mathrm{R} \rightleftharpoons \mathrm{LR} \tag{10.1}$$

In the framework of this simple model, the behaviour of an antagonist will be described successfully. But in the case of an agonist, the receptor is expected to be activated for inducing further intracellular signalling processes. So, the presented model summarizes the formation of a ligand-receptor-complex and the effect of the receptor activation resulting in the question whether the denoted complex formation will be described by one chemical equilibrium or it might be necessary to formulate a consecutive two step mechanism like:

$$\mathrm{R} \rightleftharpoons \mathrm{R}^* \tag{10.2}$$

A. Strasser, H.-J. Wittmann, *Modelling of GPCRs*,
DOI 10.1007/978-94-007-4596-4_10, © Springer Science+Business Media Dordrecht 2013

$$\mathrm{L} + \mathrm{R}^* \rightleftharpoons \mathrm{LR}^* \tag{10.3}$$

where R* denotes the activated receptor. Because it is an experimentally difficult task to distinguish between reaction (10.1) on the one hand and the reaction system (10.2) and (10.3) on the other hand, we will make the assumption that the Eq. (10.1) is also valid in the case of an agonist, where the process reads:

$$\mathrm{L} + \mathrm{R} \rightleftharpoons \mathrm{LR}^* \tag{10.4}$$

implying the receptor activation during the formation of the ligand-receptor complex.

10.3 Thermodynamic Basics

To understand the behaviour of antagonists and, with respect to the activation process, agonists, it is necessary to discuss the formation of the ligand-receptor-complex on a quantitative level. According to the first and second law of thermodynamics, the chemical behaviour of any species i is governed by its electrochemical potential $\widetilde{\mu}_i$, the chemical energy per mole, which mainly comprises in our framework of three energetic parts:

First, there is the contribution of the atomic nuclei of all the atoms present in a molecule of species i, its kinetic energy and the potential energy resulting from the chemical bonds between the atoms of the molecule. In a simple view, these contributions may be described as the chemical nature of the species i. As the reaction takes place in a solution phase, a further energetic contribution resulting from the independent interaction of each molecule of species i with the surrounding solvent has to be taken into account. All these energy terms are summarized in the quantity μ_i^o, the so-called reference chemical potential per mole of species i. Cellular systems often exhibit compartiments by means of membranes, which are not permeable for all species, particularly ions. As a consequence parts of the whole system show an electrostatic potential ϕ apart from zero which leads to a second molar potential energy contribution for a charged particle of species i characterized by its valency z_i, given by

$$z_i F \phi \tag{10.5}$$

where F denotes the Faraday constant, i.e. the charge of one mole of an arbitrary ion with valency $+1$.

The last contribution to the electrochemical potential has its origin in the second law of thermodynamics: The species i will be more stable, if a given number of moles n_i will be distributed over a larger system volume V, i.e. the molar concentration c_i will exhibit a lower value. Moreover, the displacement of the particles of each species present in the system is altered by the concentration dependent intermolecular interactions between the solute particles and between the solute and the solvent, giving rise to the so-called activity coefficient y_i of each species i. The resulting term reads as

$$RT \ln \frac{c_i y_i}{c_o} \tag{10.6}$$

where R is the gas constant, T is the absolute temperature and ln denotes the natural logarithm. Very often, the above expression is written in the form

$$RT \ln a_i \tag{10.7}$$

where a_i denotes the activity of species i in the solution. It is worth mentioning that an activity coefficient may be formulated as an analytical expression only in the case of very simple systems, for instance in the case of a solution of potassium chloride in water. Biochemical systems however show very complex interactions and therefore it is impossible to represent the activity coefficient by a simple algebraic expression. But for a dilute solution the mentioned interactions will become very small resulting in an activity coefficient near value 1. Restricting our considerations to a concentration range of about 10^{-9} to 10^{-6} M, neglecting the activity coefficient will be a good approximation and rewrites the third energy contribution as:

$$RT \ln \frac{c_i}{c_o} \tag{10.8}$$

The quantity c_o is the so-called reference concentration of 1 mol/l which results from theoretical considerations.

Summarizing all the energy terms leads to:

$$\widetilde{\mu}_i(p, T, c, \phi) = \mu_i^o(p, T) + z_i F\phi + RT \ln \frac{c_i}{c_o} \tag{10.9}$$

It should be noted that the quantities $\widetilde{\mu}_i$, μ_i^o and c_i are dependent on the system variables pressure and temperature. Generally, we are interested in ligand-receptor interactions taking place at constant pressure and temperature, so we will not use the explicit functional notation of $\widetilde{\mu}_i(p, T)$, $\mu_i^o(p, T)$ and $c_i(p, T)$. Moreover, the quantity μ_i^o does not depend on the concentration of any solute present in the solution. After discussing the fundamental thermodynamic function 10.9, we have to deal with the question of applying this concept to the ligand-receptor-interaction resulting in the chemical equilibrium according to Eq. 10.1. On the one hand we have a solution, possibly provided with an electrostatic potential ϕ, containing the ligand L and a ligand-free receptor R, embedded in its membrane and both, L and R are generally charged. The electrochemical potentials read as follows:

$$\widetilde{\mu}_L = \mu_L^o + RT \ln \frac{c_L}{c_o} + z_L F\phi \tag{10.10}$$

and

$$\widetilde{\mu}_R = \mu_R^o + RT \ln \frac{c_R}{c_o} + z_R F\phi \tag{10.11}$$

where μ_L^o is the reference potential of the ligand with respect to the solvent, i.e. water and μ_R^o is the reference potential of the receptor embedded in the membrane, but in contact with the solvent, just as the ligand L. Because of the fact that both, the ligand and the receptor are located in the same environment, the electrostatic potential acting on each other is the same. Therefore, we use the same quantity ϕ in Eqs. 10.10 and 10.11. On the other hand, we have the complex LR containing the ligand in the binding pocket of the receptor, situated in the same solvent system as the ligand and the empty receptor and therefore subjected to the same electrostatic potential ϕ. So, its electrochemical potential reads:

$$\widetilde{\mu}_{LR} = \mu_{LR}^o + RT \ln \frac{c_{LR}}{c_o} + z_{LR} F\phi \tag{10.12}$$

In case of a chemical equilibrium at constant pressure and constant temperature, the second law of thermodynamics states that the sum of the electrochemical potentials of the products, right hand side of Eq. 10.1, equals the sum of the electrochemical potentials of the educts, left hand side of Eq. 10.1:

$$\begin{aligned} &\mu_{LR}^o + RT \ln \frac{c_{LR}}{c_o} + z_{LR} F\phi = \\ &\mu_L^o + RT \ln \frac{c_L}{c_o} + z_L F\phi + \mu_R^o + RT \ln \frac{c_R}{c_o} + z_R F\phi \end{aligned} \tag{10.13}$$

Because the ligand, the receptor and the ligand-receptor-complex are charged generally, appropriate counter ions have to be present in an electrically neutral solution. For the discussion of the thermodynamics of the association process, we presuppose, that these counter ions do not influence the formation of the ligand-receptor-complex.

Due to the following equation

$$z_{LR} = z_L + z_R \tag{10.14}$$

the terms containing the electrical potential cancel and after rearrangement, we arrive at:

$$\mu_{LR}^o - \mu_L^o - \mu_R^o = -RT \ln \frac{c_{LR}/c_o}{(c_L/c_o)(c_R/c_o)} \tag{10.15}$$

where the argument of the logarithmic term equals the equilibrium constant K of the process Eq. 10.1:

$$K = \frac{c_{LR}/c_o}{(c_L/c_o)(c_R/c_o)} \tag{10.16}$$

The left hand side of Eq. 10.15 is defined as the reference Gibbs energy ΔG^o of the reaction in the case of constant pressure and constant temperature:

$$\Delta G^o = \mu_{LR}^o - \mu_L^o - \mu_R^o \tag{10.17}$$

Thus, we have the fundamental equation, valid for all types of reaction taking place at constant pressure and temperature:

$$\Delta G^o = -RT \ln K \tag{10.18}$$

Note, that in the framework of an exact thermodynamic treatment of chemical reactions, the equilibrium constant K does not exhibit any unit. Nevertheless, nearly all papers dealing with the determination of equilibrium constants of the ligand binding process provide units like nM or μM, for example, in connection with an thermodynamic equilibrium constant. This mistake results from omitting the reference concentration c_o and exhibits serious difficulties when calculating the energy quantity ΔG^o. Getting the Gibbs energy of reaction 10.1, the Eq. 10.18 is to be used, but the evaluation of the logarithm of a quantity taking a unit does not make any sense. So, how can we get the desired result in this case in an exact manner? Because c_o is defined as 1 mol/l, the given equilibrium constants have to be converted into a molar quantity and only the number is to be used in all subsequent calculations. If, for example, the dissociation constant is given as 250 nM, the quantity K in Eq. 10.18 has to be substituted by 250×10^{-9} in order to get the true value of ΔG^o for the ligand-receptor-dissociation process. The quantity ΔG^o comprises two terms, firstly, the enthalpy ΔH^o and secondly the entropy ΔS^o,

$$\Delta G^o = \Delta H^o - T \Delta S^o \tag{10.19}$$

where T represents the temperature and

$$\Delta H^o = H^o_{LR} - H^o_L - H^o_R \tag{10.20}$$

$$\Delta S^o = S^o_{LR} - S^o_L - S^o_R \tag{10.21}$$

With respect to the chemical reaction Eq. 10.1 the quantities ΔG^o, ΔH^o and ΔS^o apply to the formation of one mole of ligand-receptor complex LR. The enthalpy term ΔH^o contains information about the change in energy during a particular reaction, whereas the entropy term ΔS^o lacks any simple interpretation. Nevertheless, very often ΔS^o is connected with the concept of order and disorder in the course of chemical reactions. As these terms are not defined exactly, the interpretation of ΔS^o in most cases leads to severe mistakes in inspecting of chemical processes and should therefore used with caution.

To gain more detailed insight in the ligand-receptor interaction, that is to understand the magnitude of the equilibrium constant K, we have to analyze the energy term ΔH^o and the entropy term ΔS^o on a molecular level. But before doing so, we first have to deal with the determination of these two quantities, which will be the subject of the following section.

10.4 Evaluating ΔH^o and ΔS^o

One of the methods to determine ΔH^o is given by the isothermal titration calorimetry. To elucidate the basic principles of this method, we will consider a solution containing the receptor R with a certain concentration c_R^o. Assuming constant temperature and constant pressure, we will add a small amount of a stock solution of the ligand L in a way that its actual concentration in the titrand solution is c_L^o. The reaction, Eq. 10.1, will take place and an equivalent portion of the ligand and receptor will form the complex LR, with a concentration dependent on the up to now unknown equilibrium constant *K*. Because of the chemical process, an enthalpy change will occur which we are able to determine by the mentioned calorimetric method. Assuming the receptor concentration is small compared to the ligand concentration and neglecting the increase in the system volume caused by adding the stock solution, the concentration of the free ligand nearly remains constant and the amount of complex formed after establishing the chemical equilibrium is given by the relation:

$$K = \frac{c_{LR} c_o}{c_L^o \left(c_R^o - c_{LR} \right)} \tag{10.22}$$

where *K* is the unknown equilibrium constant and the term $c_R^o - c_{LR}$ denotes the concentration of the free receptor c_R. Solving Eq. 10.22 for c_{LR} yields the following result:

$$c_{LR} = K \frac{c_L^o}{c_o} \frac{c_R^o}{1 + K \frac{c_L^o}{c_o}} \tag{10.23}$$

The corresponding change in enthalpy Δh per unit volume for dilute solutions is then given by:

$$\Delta h = c_{LR} \Delta H^o \tag{10.24}$$

where, in the case of dilute solutions ΔH^o approximately does not depend on the concentration of any reactant. Substituting c_{LR} from Eq. 10.23 into 10.24 denotes the quantity Δh as function of the total ligand concentration c_L^o caused by adding successive amounts of the stock solution containing the ligand species:

$$\Delta h = \Delta H^o K \frac{c_L^o}{c_o} c_R^o \left(1 + K \frac{c_L^o}{c_o} \right)^{-1} \tag{10.25}$$

By applying a nonlinear least square fit method, we determine the unknown parameters ΔH^o and *K*. Knowing about *K*, we are able to calculate ΔG^o according to Eq. 10.18 and with the help of Eq. 10.19, we get ΔS^o.

Another method uses the temperature dependence of the equilibrium constant and the related quantity ΔG^o to determine ΔH^o and ΔS^o. Starting with Eqs. 10.18 and 10.19, we can write

$$\Delta H^o - T \Delta S^o = -RT \ln K \tag{10.26}$$

The measurement of K at the temperature of interest leads to an equation, containing two unknowns, ΔH^o and ΔS^o for that temperature. To solve this problem, we could assume temperature independent quantities ΔH^o and ΔS^o, so the determination of K at a series of temperatures would lead to a linear relationship between T and the right hand side of Eq. 10.26 with slope $-\Delta S^o$ and intersection ΔH^o. But extensive investigations of the association constant at different temperatures reveal a distinctive dependence of ΔH^o and ΔS^o on temperature. Thus, the above mentioned linear relationship between ΔG^o and T will no longer hold. To overcome this difficulty, we make use of the fundamental thermodynamic relations in the case of constant pressure:

$$\frac{\partial \Delta H^o}{\partial T} = \Delta C_p^o \tag{10.27}$$

$$\frac{\partial \Delta S^o}{\partial T} = \frac{\Delta C_p^o}{T} \tag{10.28}$$

where the quantity ΔC_p^o denotes the change in the heat capacity for the reference state during the reaction and reads:

$$\Delta C_p^o = C_{p,LR}^o - C_{p,L}^o - C_{p,R}^o \tag{10.29}$$

$C_{p,LR}^o$, $C_{p,L}^o$ and $C_{p,R}^o$ denote the heat capacity of the complex, the ligand and the receptor within the solution in its particular reference state. These terms are independent of concentration, but are generally functions of the temperature and the pressure. If we are interested in the values of ΔH^o and ΔS^o at a given temperature T_o, we can determine the association constants K at a series of temperatures, enclosing T_o, by evaluating the particular concentration of the ligand-receptor complex at given c_R^o and c_L^o with the help of radioligand competition binding assays (Weiland et al. 1979; Wittmann et al. 2009). Here, we also assume that c_L^o is much larger than c_R^o and so we are able to calculate ΔG^o for each temperature according to Eq. 10.18. To combine ΔH^o and ΔS^o with ΔG^o at different temperatures, Eq. 10.19, we integrate Eqs. 10.27 and 10.28 in the range from T_o to any temperature T of the series. If the temperature interval of the measurement is small, for example 25 K, it is a good first approximation to think of ΔC_p^o as a constant for a given reaction according to Eq. 10.1 and the integration of Eqs. 10.27 and 10.28 yields:

$$\Delta H^o(T) = \Delta H^o(T_o) + \Delta C_p^o(T - T_o) \tag{10.30}$$

$$\Delta S^o(T) = \Delta S^o(T_o) + \Delta C_p^o \ln \frac{T}{T_o} \tag{10.31}$$

Substituting these results into Eq. 10.18, we arrive at:

$$\Delta H^o(T_o) + \Delta C_p^o(T - T_o) - T\left(\Delta S^o(T_o) + \Delta C_p^o \ln \frac{T}{T_o}\right) = -RT \ln K \tag{10.32}$$

Having determined the association constant K at various temperatures T, a linear least square fit algorithm enables us to calculate ΔH^o, ΔS^o and ΔC_p^o at the temperature T_o, which is commonly defined as 298.15 K.

10.5 Special Topics

Within this section, we will discuss the possibility that a ligand is able to bind in more distinct orientations inside the receptor (Strasser et al. 2010a). For the sake of simplicity, we will restrict our considerations on the case of two different orientations and consequently have to define two association processes:

$$\mathrm{L} + \mathrm{R} \rightleftharpoons \mathrm{LR1} \tag{10.33}$$

and

$$\mathrm{L} + \mathrm{R} \rightleftharpoons \mathrm{LR2} \tag{10.34}$$

where L and R on the left hand sides of the Eqs. 10.33 and 10.34 correspond to exactly the same compounds, whereas LR1 and LR2 denote two distinct ligand-receptor complexes. Making use of the assumptions within this chapter, the equilibrium constants for these reactions are given in accordance to Eq. 10.16 by:

$$K_1 = \frac{c_{LR1} c_o}{c_L^o (c_R^o - c_{LR1} - c_{LR2})} \tag{10.35}$$

and

$$K_2 = \frac{c_{LR2} c_o}{c_L^o (c_R^o - c_{LR1} - c_{LR2})} \tag{10.36}$$

Appropriate experimental techniques enable us to determine the concentration of exactly one complex LR. But the same methods applied onto the case of two ligand-receptor complexes will result in the determination of the sum

$$c_{LR1} + c_{LR2} \tag{10.37}$$

which reads as:

$$c_{LR1} + c_{LR2} = (K_1 + K_2) \frac{c_L^o}{c_o} \frac{c_R^o}{1 + (K_1 + K_2) \frac{c_L^o}{c_o}} \tag{10.38}$$

The right hand side of Eq. 10.38 exhibits the same form as the right hand side of Eq. 10.23 in the case of exact one ligand-receptor complex. As a consequence, it is impossible to determine the binding constants K_1 and K_2 of each complex separately, using traditional experimental techniques, but we will get only the sum $K_1 + K_2$. The only possibility to get information on the properties of the distinct orientations of the ligand in the binding pocket of the receptor exists in constructing a model for each ligand-receptor complex and calculating the corresponding binding constants afterwards. The sum of these quantities has to be compared to the experimental value of $K_1 + K_2$ for validation of the model. So, if we are forced to deal with more than two orientations, we will encounter much more difficulties to gain information about the binding properties of ligands.

Chapter 11
Important UNIX/LINUX Commands

11.1 Some Basic Aspects of the Operating System UNIX/LINUX

UNIX, especially its implementation LINUX, is a very powerful tool to perform all the tasks in the framework of molecular modelling On the one hand, a lot of programs dealing with molecular modelling make use of the operating system UNIX, which on the other hand offers an extensive set of commands for an effective manipulation of data files necessary for different runs of modeling programs. The central interface for getting the benefit of this performance is the so-called "shell", the command line interpreter of a UNIX system. In the last decade, many different shells have become available to the user. For example the Bourne-Shell (sh or bash), the Korn-Shell (ksh), the C-Shell (csh) or the TC-Shell (tcsh), which will be utilized in this book. The syntax of most, so-called external, UNIX-commands is independent of the choice of the shell. Differences appear in commands internal to the shell or when one makes use of the meta-characters. Thus, when a shell other than tcsh is used, it is advisable to contact the appropriate manual pages to get information about specific syntax elements whereas most of the features of the tcsh will also be valid in the csh command interpreter. To reproduce the examples and exercises of the following chapters, the reader should be familiar with the basic concepts of files and directories, including the concept of access modes and the corresponding commands to create, remove, copy, rename or list these objects. Furthermore, the user should be able to operate on a text editor, like "`gedit`" or "`vi`".

11.2 The Use of Shell Operators and Meta-Characters in Tcsh Environments

A UNIX command line exhibits the following general structure:

```
command options objects
```

Therein, the "command" denotes a particular UNIX command possibly completed by "options" which control its execution or output. The item "objects" indicate files

A. Strasser, H.-J. Wittmann, *Modelling of GPCRs*,

DOI 10.1007/978-94-007-4596-4_11, © Springer Science+Business Media Dordrecht 2013

and/or directories. Note that all parts of the command line must be separated by at least one white space character, e.g. a blank. In the following sections each command line is introduced by the so-called shell prompt, indicated by the character ">". The "ENTER"-key, specifying the end of the command, is given by the symbol " ↵".

To find out the contents of a directory named "`dir`", located in the working directory, one would us the command `ls`:

```
> ls dir ↵
```

which generally brings up a very poor listing whereas the command

```
>ls -al dir ↵
```

prints out an extensive list of objects including a lot of its properties, caused by the option "`al`", which is indicated by a hyphen.

Shell commands may be combined with the symbol | to form a pipleline, for instance,

```
> echo "abc" | tr 'a-z' 'A-Z' ↵
```

will translate all lower case characters of the string "`abc`" into upper case characters, will say, the output of the command on the left hand side of the pipe-symbol | is used as input for the command on the right hand side (for a description of `tr` refer to the following section). An arbitrary number of commands may be combined in this manner.

Another group of operators consists of the symbols $<$, $>$, $\gg$ known as file input/output redirection. Suppose a program named "pgm" which normally reads from the standard input; providing a file named "`data`", "pgm" may get its input from this file instead from standard input:

```
>pgm < data ↵
```

To redirect output of a command to a file use the operator ">":

```
> ls -al dir > temp ↵
```

The output of the command `ls` will be stored in a file named "`temp`". Note that an already existing file "`temp`" will be destroyed before redirecting output! To append the output of a command to an existing file named "`data`", use the symbol >>:

```
> ls -al dir >> data ↵
```

Additionally, a command line may include meta-characters which will be interpreted by the shell in a special manner, also known as shell substitution. Thus, the original command line will be altered after that.

11.3 Shell Substitutions

The following section will deal with the most important shell substitutions.

11.3.1 File Name Substitution

If a string contains any of the characters *, ?, [] or {}, the file name substitution occurs.

For instance, to remove all objects from the current working directory, we use the command

```
> rm -r * ↵
```

Note, in a basic UNIX environment, the user is not prompted before removal and all files and directories will be lost with two exceptions: If an object name begins with a period or contains a slash /, the meta-character * does not affect them. In this example, we used the special character * to efficiently address the contents of the current working directory

```
> ls -al * ↵
```

The behaviour of the shell is as follows: The character * is substituted by all entries of the working directory and afterwards, the shell executes the command.

Now, lets have a look on the directory "data" located in `/usr/project` with a couple of files:

`dat1`, `data.old`, `data.save`, `dat1.new`, `dat2.new`, `dat3.new`, `dat3.old`, `geo.new`.

Assume, a user might want to move the files named `data.old` and `data.save` to a directory `/usr/new_project`. Making use of the meta-character *, the command would look like this:

```
> mv /usr/project/data/data* /usr/project_new ↵
```

Thus, the character * substitutes an arbitrary string including the null-string.

Another special character ? matches exactly one character, so the command

```
> mv /usr/project/data/dat?.new /usr/project ↵
```

would move the files `dat1.new`, `dat2.new` and `dat3.new` to the directory `/usr/project`. The notation

```
> mv /usr/project/data/dat[13].* /usr/project ↵
```

moves the files `dat1.new`, `dat3.new` and `dat3.old` to the directory `/usr/project`. Thus, an enumeration of characters enclosed in square brackets will match a single character out of this enumeration. An extension of the meta character [...] is given by the pattern {...}. The meta-notation `{dat1,geo}.new` expands to `dat1.new` and `geo.new`.

11.3.2 Variable Substitution

Use of shell variables within commands will make work easier and more efficient. The command `set` allows a user to define a name and assign a value to it:

```
> set var = 123 ↵
```

declares a variable var with the value 123.

The name of a variable consists of case-sensitive letters, digits and the underscore (_) not starting with a digit. To reference the value of a variable, the meta-character $ is used. Thus, the command

```
> echo $var ↵
```

will print "123" to standard output. To remove a variable, use the unset command:

```
> unset var ↵
```

This will destroy the variable var.

Assume, we will frequently copy files from a directory named /share/data/project/md. Defining a variable dir with the value of the mentioned object

```
> set dir = /share/data/project/md ↵
```

will simplify for example the cp-command:

```
> cp $dir/enzyme* . ↵
```

Here we make use of the two meta-characters $ for variable substitution and * for file name substitution. So, all files in /share/data/project/md whose names start with enzyme will be copied to the current working directory, indicated by a dot.

Now, define a variable x with value 123

```
> set x = 123 ↵
```

The command

```
> echo $xabc ↵
```

will result in an error message, indicating, that the shell will not recognize the name xabc. To prevent the shell from misinterpreting the string xabc surround the name x by braces:

```
> echo ${x}abc ↵
```

This command will print 123abc to the standard output. Here, the use of {...} does not mean file name substitution rather insulating the variable name from following characters because of the special symbol $.

The symbol @ allows arithmetic (positive integer range) calculations performed with variables, e.g.

```
> set x = 1 ↵
> @ x = $x + 3 ↵
> echo $x ↵
```

Therein, the first command defines a variable x with value 1. In the next statement, the number 3 is added and finally, the new value of the variable x is printed.

11.3.3 Command Substitution

If the shell encounters a string enclosed in back quotes, the command substitution takes place. The string is considered as a command and is executed in a subshell. Its output replaces the string including the back quotes in the original command.

Have a look onto the sequence of DNA bases `ATCctgcgtatccccCCT` which is to be checked for an even number of triples made of lower case characters. For this we can make use of the command `expr` (for a complete description see the appropriate manual page) to evaluate the arithmetic remainder of a number divided by 3 for example:

```
> expr 21 % 3 ↵
```

will print out zero. But how to determine the number of lower case characters in the sequence? First, we will define a variable named `base` to hold the complete sequence:

```
> set base = ATCctgcgtatccccCCT ↵
```

Now we are able to extract the substring consisting of lower case characters and evaluate its length with the help of the following pipeline:

```
> echo -n $base | tr -d 'A-Z' | wc -m ↵
```

where the echo command is used with the option –n to suppress the trailing newline character which erroneously would be counted by `wc` (see command section). The `tr` command deletes all upper case characters of the string. Enclosing the command line above in back quotes as the first argument, the command `expr` outputs the desired result:

```
> expr 'echo -n $base | tr -d 'A-Z' |
  wc -m` % 3 ↵
```

Note, the use of a variable substitution as part of a command substitution!

11.3.4 Protection Mechanism for Meta-Characters of the TC-Shell

Sometimes it is necessary to suppress part of the shell substitution or to have meta-characters as valid characters not modifying the original command line. For this purpose, tcsh provides the backslash or single quotes and double quotes: A character preceded by a backslash, for example *, will not be expanded. Note that the backslash is needed to prevent the shell from a special treatment of the symbol `!`. To protect more than one meta-character from shell substitution within a string, single quotes might be used.

Suppose, we have defined the variable `x` with value `123` via

```
> set x = 123 ↵
```

then the commands

```
> echo abc\$x ↵
```

and

```
> echo 'abc$x' ↵
```

will both write the string `abc$x` to the standard output. As another example, the command line

```
> echo '*$x' ↵
```

would not be expanded in the sense of filename and variable substitution but simply outputs the string `*$x`. Strings enclosed in double quotes will still be command and variable expanded. Thus, the result of the command

```
> echo "*$x" ↵
```

will have `*123` as its output.

11.4 Discussion of Selected LINUX Commands

The following section lists some important LINUX commands for processing ASCII (i.e. human readable) files. Each command section is divided into three subsections **syntax**, **explanation** and **example**.

The **syntax** subsection only mentions the most relevant instances of the command. For a complete description the reader is encouraged to consult the corresponding LINUX manual page. The **explanation** subsection gives some more information of the command, which is finally discussed in the **example** section with the help of simple exercises. In most cases, the contents of a file, which will be created later on within an exercise, related to the command `cat` (see below), will be processed. To reproduce the following examples, the reader is supposed to have opened a shell-terminal, primarily a tcsh shell-terminal. For the use of a bash shell-terminal, one has to take into account a different meaning of the shell meta-characters.

`cat`

syntax

```
cat file
cat > file
```

explanation

The first form prints the contents of a file to standard output, whereas the second form may be used to create a simple data file.

example

Create a file in the so called csv-format with the name "`data`" using the second form of the command:

```
> cat > data ↵
```

Then, the text cursor will be placed on the beginning of the next line. Now enter the following strings, each terminated by a newline character:

```
1;DRG;3,39;2952;24,80;
2;DRG;3,42;2934;24,92;
3;DRN;3,29;3043;24,37;
4;DRG;2,29;4376;24,46;
5;SOL;2,13;4719;24,75;
6;UNK;2,06;4864;24,74;
```

To finish data input press the buttons „Strg“ (or „Ctrl“) and „d“ (abbreviation ^d) simultaneously at the beginning of a new line to signalize END OF FILE to the shell.
Now, the command

```
> cat data ↵
```

will print out the contents of the recently created data file with the name “data”.

`cut`

syntax

```
cut -c n file
cut -c m-n file
cut -c m,n,... file

cut -d 'delim' -f n file
cut -d 'delim' -f m,n,... file
cut -d 'delim' -f m-n file
```

explanation

The first three instances of `cut` will perform the following tasks:

- Print out the n'th character of each line of a file
- Print out a range of characters m to n of each line of a file
- Print out the characters m, n, ... of each line of a file

The last three commands use a character “delim” (enclosed in single quotes) to divide each line of a file into fields. A line beginning with the character “delim” forces an empty first field. Field numbering always starts with one. On output, the specified fields will be separated by the character “delim”. Thus, the last three commands will lead to the following results:

- Print out field n of each line of a file using the character “delim” as delimiter

- Print out field m to n of each line of a file using "delim" as delimiter
- Print out the fields m, n, ... of each line of a file using the character "delim" as delimiter

example

Write out character 7 of each line of file `data`

`> cut -c 7 data ↵`

Write out characters 7 to 9 of each line of file `data`

`> cut -c 7-9 data ↵`

Write out characters 2,4,6,8 of each line of file `data`

`> cut -c 2,4,6,8 data ↵`

Print field number 3 using the delimiter "," of each line of the file `data`

`> cut -d ',' -f 3 data ↵`

Write out the first, second and fourth field of each line of the file `data` using "`;`" as delimiter

`> cut -d ';' -f 1,2,4 data ↵`

Finally generate an output with the fields 2–4 of each line of file data, using the delimiter "`;`"

`> cut -d ';' -f 2-4 data ↵`

`gawk`

syntax

`gawk 'pattern{actions} ...' file(s)`

explanation

`gawk` (sometimes referred to as "`awk`") certainly is the most powerful command for UNIX. In its simplest form it consists of a sequence of pattern-action statements. Each input line of file(s) matching the pattern is divided into fields using blanks and/or tabs as delimiters. The value of each field is referenced by the notation "`$#`" where # denotes the field number starting at 1. The special notation "`$0`" refers to the entire line of an input file. Actions without any patterns will be performed for each input line. As special cases the patterns "`BEGIN`" and "`END`" mark actions which will be performed before reading the first line of input and after the last input line has been processed. Because of the versatility of this command, the reader is strongly recommended to contact the manual page for a complete description of `gawk`, so the next section will contain only some simple applications. Extensive examples making use of gawk may be found in Chap. 7 and Sect. 11.7.

example

Given the file `data` containing information about a very large molecule, one line per atom, we want to print the x-, y- and z-coordinates of some atoms located in columns 6, 8 and 10 beginning at line 2500 and ending at line 3456. The appropriate values will be stored in a file named `outdat`.

The command would look like this:

```
> gawk 'NR>= 2500 && NR<= 3456 {print
$6,$8,$10}' data > outdat ↵
```

The pattern "`NR>=2500 && NR <=3456`" uses the built-in variable `NR` which holds the actual line number. The notations ">=" and "<=" represent the relational operators "greater than" and "less than" whereas the symbol "`&&`" means the locial AND operator. Gawk reads one line after the other from the input file `data` and prints out the field numbers 6, 8 and 10 only from the lines in the range between 2500 and 3456, inclusive.

Assume there is an output file `out` from a simulation run, which contains a line holding the heat of formation in the form:

```
HEAT OF FORMATION = −1345.774 kJ/mol
```

To get the value 1345.774 in Joule per mole, the appropriate line, indicated by the term "`HEAT OF`", has to be located and the value of field number 5 multiplied by thousand has to be printed out. Take into consideration that any number of repeated blanks and/or tabs count as a single delimiter, where leading delimiters will be ignored. Thus, the command will read:

```
> gawk '/HEAT OF/ {print $5*1.0e3}' out ↵
```

In this case, the pattern consists of a regular expression "`HEAT OF`" enclosed in slashes indicating the line of the file `out` which will be processed by `gawk`.

`grep`

syntax

```
grep 'regular expression' file
grep -v 'regular expression' file
grep -c 'regular expression' file
grep -n 'regular expression' file
```

explanation

The four command lines given above, will lead to the following results:

- `grep` searches the named file for lines containing a mach to the specified regular expression and writes them to standard output. For a complete description of regular expression used by `grep` see the appropriate manual page.
- The second command inverts the search, i.e. the output comprises of all lines not matching the regular expression.
- The third instance of `grep` prints the number of matching lines for file.
- The last command form prefixes each line on output with its corresponding line number within the file, followed by a colon

example

Given the file `data`, created by the command cat in the appropriate section, we want to extract all lines containing the pattern "`DRG`" using grep in its first form:

```
>grep 'DRG' data ↵
```

where the corresponding lines may contain the pattern "`DRG`" in any position; to specify this position more exactly would lead to the following statement:

Find all lines where the string "`DRG`" is located after a line number followed by the character "`;`". In this case, the regular expression and the corresponding command would look like this:

```
> grep '^[0-9]\+;DRG' data ↵
```

The character ^ specifies the beginning of a line and the sequence `[0-9]\+` means a number of the set 0–9, a pair of square brackets denotes a set, repeated one or more times, indicated by the character `+`, where the plus sign must be preceded by a backslash to signalize the special meaning "one or more". All other characters stand for themselves.

To print out all lines of the file data, not containing the string `DRG` use

```
> grep -v 'DRG' data ↵
```

To print only a count of all lines containing for example the pattern "`29`", type:

```
> grep -c '29' data ↵
```

The command

```
> grep -n '29' data ↵
```

preceeds each line containing the pattern "`29`" by its line number.

Now, let us have a closer look to the command `grep -c '29' data`: We get all the lines containing the regular expression `'29'` anywhere. But how to solve the problem of counting all the lines containing the pattern '29' in the fourth field of a line, assuming the character : as a delimiter? Remember the command `cut` in the form

```
> cut -d ';' -f 4 data ↵
```

which will output the fourth field of each line of data. Combining this command with the help of the so called pipe symbol | with the command `grep`, using the regular expression '29' will show the desired result:

```
> cut -d ';' -f 4 data | grep -c '29' ↵
```

`head`

syntax

```
head -n number file
```

explanation

Print out line one up to number of the specified file.

example

Print the first three lines of the file "data"

```
> head -n 3 data ↵
```

`sed`

syntax

```
sed 's/pattern1/pattern2/' file
sed 'ms/pattern1/pattern2/' file
sed 'm,ns/pattern1/pattern2/' file
sed -n 'm,np' file
```

explanation

The first form of the command `sed` replaces the sequence of characters in `pattern1` with the sequence of characters in `pattern2` once for each input line of `file`.

The second and third form of the command `sed` will do the replacement only for line `m` and for lines `m` to `n` respectively of the named `file`.

If there are multiple instances of `pattern1` to be replaced by `pattern2` on one input line, the character `g` has to be appended after the last slash of the quoted part, for instance:

```
sed 'ms/pattern1/pattern2/g' file
```

The first three `sed` command print all the modified and unchanged lines to the standard output. The last form of the command just writes lines `m` to `n` of `file` to standard output.

example

First, substitute the string `DRG` by `DRN` in all lines of the file `data`:

```
> sed 's/DRG/DRN/' data ↵
```

Do the same only for the line number 4

```
> sed '4s/DRG/DRN/' data ↵
```

In the next example, change the string `DRG` in line one and two to `UNK`:

```
> sed '1,2s/DRG/UNK/' data ↵
```

To print out lines 2 up to 5 of the file `data`, use:

```
> sed -n '2,5p' data ↵
```

`tail`

syntax

```
tail -n m file
tail -n +m file
```

explanation
The first form will print out the last `m` lines of `file`, whereas the second form will print lines beginning with number `m` to the end of `file`.

example
Write out the last four lines of the file "`data`"

```
> tail -n 4 data ↵
```

Write out all lines of file data beginning with line number 4

```
> tail -n +4 data ↵
```

`tr`

syntax

```
tr 'pattern1' 'pattern2'
tr -s 'pattern'
tr -d 'pattern'
```

explanation

`tr` is a filter command receiving its input for example by a pipe and prints results to standard output. `pattern`, `pattern1` and `pattern2` each represent a sequence of characters. The number of characters in the sets `pattern1` and `pattern2` should be equal.
Assuming an input file named data using the following command

```
cat data | tr 'pattern1' 'pattern2'
```

translates each character of the set `pattern1` into the corresponding character of `pattern2` for all the lines of the file data.

The command

```
cat data | tr -s 'pattern'
```

replaces each sequence of repeated characters listed in pattern with a single occurrence of that character for each input line of the file data.

The command

```
tr -d 'pattern'
```

deletes the characters listened in `pattern` from each input line.

example

The file `data` contains items in the so called csv format, e.g. the character "`;`" separates the data fields and the decimal point is replaced by a comma. Most LINUX commands require fields separated by one or more spaces or tabs and expect a decimal point. Thus, one has to transform the contents of data with help of the command `tr`:

```
> cat data | tr ';,' ' .' ↵
```

Have a look onto the string "`aBBcAAAfrBB`", where the repeated characters "`A`" and "`B`" are to be replaced by a single character "`A`" and "`B`" in order to yield the sequence "`aBcAfrB`":

```
> echo ´aBBcAAAfrBB´ | tr -s 'AB' ↵
```

Now, assuming the field delimiter "`;`", we will delete the sequence of upper case characters in the second field in all the lines of the file `data`, using the pattern range '`A-Z`':

```
> cat data | tr -d 'A-Z' ↵
```

As a consequence of removing the characters in each data line, the output exhibits two adjactent "`;`" characters which are to replaced by a single "`;`" character with the help of a further instance of the `tr` command:

```
> cat data | tr -d 'A-Z' | tr -s ';' ↵
```

`wc`

syntax

```
wc -l file
wc -m file
```

explanation

In its first form, the command `wc` writes out the number of lines of `file`. The second form of the `wc` command counts the characters within `file`.

example

Count the lines of the file `data`:

```
> wc -l data ↵
```

An example of `wc`, using the option `-m` for counting characters is already presented in the section "Command Substitution" above.

11.5 Loops Statements of the Tcsh Shell

Loops are very helpful in solving problems by combining arbitrary commands and executing them repeatedly. Two loop constructs are available in the tcsh shell. First, we mention the `foreach` loop.

foreach-loop

syntax

```
foreach variable (value1 value2 ...)
    command1
    command2
    ...
    ...
end
```

explanation and examples

The command sequence is executed for each of the values `value1`, `value2, ...` . Afterwards the loop exits.

Suppose, we got a lot of data files. Each of them is named with the starting character x. To save these files, we will rename them enclosing their names in #...#. Note, that a command like

```
mv x* #x*#
```

will not work because the shell will not distinguish between source and destination files. Thus, we will make use of the foreach loop:

```
> foreach i (x*) ↵
    > mv $i #$i# ↵
> end ↵
```

The `foreach` statement defines a variable i and substitutes for x* all file names beginning with x in the current working directory. The loop statement, say the `mv` command, now moves one file after the other with the help of a variable substitution. Finally, the loop statements are finished by the `end` statement.

A second possibility to form a loop is realized by the while loop:

while-loop

syntax

```
while (expression)
    command1
    command2
    ...
    ...
    ...
end
```

explanation and examples:

The `while` expression is evaluated and has value 1 if it is of arithmetic type with value not equal to zero or if it is an expression which evaluates to true whereas in all other cases expression has a value zero. Now the command sequence is executed as long as the `while` expression has the value 1. The user has to provide a command altering the value of this expression in order to leave the loop after a certain number of runs.

Given the following sequence of DNA bases which is stored in a variable, named `base`:

```
> set base = atgtctttcctcccaggaatgacc ↵
```

After testing that the remainder of the number of characters in `base`, indicated by the `%` character, divided by three, yields zero

```
> expr $%base % 3 ↵
```

we are going to split this sequence into triples and write them to standard output. First we define a variable `n` holding the number of base characters and a loop variable named `i` is initialized with 1. To create the first triple, we cut the characters with number `i` to `i+2` (represented by variable `j`) and print them out. Afterwards, the value `i` is incremented by three and we proceed as long as `i` is less than `n`:

```
> set n = $%base ↵
> set i = 1 ↵

> while ( $i < $n ) ↵
   > @ j = $i + 2 ↵
   > echo $base | cut -c $i-$j ↵
   > @ i = $i + 3 ↵
> end ↵
```

Note the incrementation of variable `i` in the command just before the end statement assures the while loop to be exited, if `i` is equal or greater than `n`.

11.6 Working with Shell Scripts

Referring to our last exercise we recognize two disadvantages: Firstly, we have to do a lot of work prior to the while statement. For each new problem, we have to repeat the steps dealing with creating shell variables and test the number of characters via `expr`. After that we have to execute all the statements of the while loop. Secondly, a mistyping within the while block makes it necessary to abort writing and to repeat all the loop statements. Making use of a so called shell script represents a possibility to avoid all these difficulties. A shell script is an ASCII file containing all commands necessary to solve a particular problem just as you would enter these commands within a shell terminal. Syntax errors would be easily eliminated in a test run and further on, the script can be applied to similar applications of a project by means of minor changes. To elucidate the implementation of a shell script we will repeat our last example. First, start an editor, for example `gedit`. Then enter the following statements, one per line. Empty lines will be ignored by the shell.

```
#! /bin/tcsh

# Determine triples of DNA base sequence

if ( $1 == "") then
    echo "*** Missing base string ***"
    echo "SYNTAX: $0 <base string>:"
    exit 1
endif

if (`expr $%1 % 3` != 0) then
    echo "*** wrong number of characters in
    base sequence ***"
    exit 1
endif

set n = $%1

set i = 1

while ( $i <= $n )
    @ j = $i + 2
    echo $1 | cut -c $i-$j
    @ i = $i + 3
end
```

After the end statement of the while loop save this file for example as `triple` and quit the editor. To start this script, the user must have appropriate rights to execute the commands within the file. Suppose, a user `mike` has created this shell script. After calling

```
> ls -al triple ↵
```

the user will get the output presented in the next line:

```
-rw-r--r-- 1 mike users 308 22. Jan 17:17 triple
```

where the columns 2–10 represent the file access mode for user `mike` (`rw-`), the owner of `triple`, the members of the group users (`r--`) and for others (`r--`), indicating that user `mike` is able to read from and write to file `triple`, whereas all other users may only read from `triple`. For `mike`, to execute a program or shell script, his access mode list should look like `rwx`. To change and verify the mode of the file triple, enter the following command in a shell terminal:

```
> chmod u+x triple ↵
> ls -al triple ↵
```

Note that the file access mode for user `mike` now has changed to `rwx`. Thus, `mike` is able to start the script by typing

```
> triple "atgtctttcctcccaggaatgacc" ↵
```

where the string not necessarily has to be quoted.

Now, have a closer look to the shell script: The first line "`#! /bin/tcsh`" indicates that a tcsh shell is to be started to execute the commands of the shell script. Scripts should always start with a first line defining the command processor. A # character in a line other than the first, introduces a comment, which extends to the end of the line. Note, the variable substitution "`$1`" which is special to shell scripts, refers to the first argument when calling `triple`. Further parameters can be referenced by `$2`, `$3` and so on. The notation `$*` means all arguments given to the command. So, the `if`-statement checks, whether an argument, the base string, is available and if not, indicated by the null string "", provides some messages to the user. Here, the notation "`$0`" refers to the command itself. After that, the script exits, caused by the command `exit`, completed by an arbitrarily error number in the range 0–255. In most cases, the exit number is of no importance to further work. The second if statement carries out a test on the right number of base triples and the script exits if it fails. All other commands correspond to statements in our loop exercise.

In order to test a few situations of calling the script, enter the following commands and have a look onto the output:

```
> triple ↵
```

```
> triple aggt ↵
```

It should be mentioned, that the script triple is subject of further extensions as to test if the base string contains for instance numbers or other special symbol. Thus, here we present only a basic work-out.

11.7 A More Extensive Example

Construction of a sequence of amino acids from a base sequence using gawk

To elucidate the use of the `gawk` command we will construct the sequence of amino acids resulting from a base sequence. First, create a file `base.dat` which holds the base sequence:

```
> cat > base.dat ↵
```

```
> ATGGCCatgtctttcctcCACCATccccct ↵
> ^d ↵
```

Next, provide a file code.dat, containing the assignment of base triplets and amino acids (see appendix 13.2):

```
> cat > code.dat ↵
> atg Met ↵
> aga Arg ↵
> gga Gly ↵
> tct Ser ↵
> ttc Phe ↵
> gca Ala ↵
> ctc Leu ↵
> ^d ↵
```

Assume, we want to translate the base sequence given by the first substring composed of lower case letters into a sequence of corresponding amino acids. We will read the file `code.dat` and create an array `AS` using the base triplets as an index string and the corresponding amino acid name as the value:

`AS[agt]` results in Met

and so on.

Let us test this part of our example:

```
> gawk '{AS[$1]=$2} END{base="aga"; print AS[base]}'
code.dat ↵
```

will output the string "Arg".

The first pattern-action statement merely consists of an action, which means that this part will be applied to all lines of the input file `code.dat`. For each line the first field denoted by "`$1`" is used as an index of the array `AS`, the second field, "`$2`" represents the value of the array element. It should be taken into account that the entries in the first column of the file `code.dat` should be different for use as array indices. The second pattern-action statement exhibits the special pattern "`END`", so the corresponding actions, separated by a semi-colon will be executed after all lines of the file `code.dat` have been processed. The first command assigns the value "aga" to string variable "base". The next statement prints out the array element `AS[base]`, which equals the amino acid arginine.

Next we will read the base alignment and extract the first substring containing only lower case letters. This part of the problem may be treated in the following way:

```
> gawk '{match($0,/[a-z]+/);
seq=substr($0,RSTART, RLENGTH)} END{print
seq,RSTART,RLENGHT}' base.dat ↵
```

The first action statement makes a call to `gawk`'s built-in-function `match`, which takes a string, here the entire line (`$0`) of the base sequence read from the file base.dat and searches for the beginning of the regular expression, enclosed in slashes, represented by the second argument of the function match. The notation "[a-z]+" means one or more occurrences of lower case characters. The second assignment defines the variable seq which holds the substring of the entire line ($0) starting at RSTART and spawning RLENGTH characters. In this case, we make use of two built-in-variables of gawk, RSTART and RLENGTH, which are set by the function match. The pattern-action statement denoted by END prints out the interesting sequence of bases, the starting position of the substring and its number of characters.

Now we will combine both gawk commands to construct the associative array AS and to extract the appropriate base sequence. The latter task will be done in the action part of the pattern BEGIN whereas the array AS is to be constructed in the action part applied to each input line of code.dat. The command will now look like this:

```
> gawk 'BEGIN{getline base <"base.dat"; match
(base,/[a-z]+/); seq=substr(base,RSTART,RLENGTH)}
{AS[$1]=$2; print AS[$1]} END{print seq}' code.dat ↵
```

Because we have two input files, base.dat and code.dat, the BEGIN action part uses the getline statement to store the contents of base.dat in the variable base, which in turn will be processed by the functions match and substr to generate the base sequence. The following action-statement will be applied to each line of the file code.dat building up the array AS, whose elements will be printed. Finally, the END action prints out the base sequence.

The complete form of the gawk command is presented next:

```
> gawk 'BEGIN{getline base < "base.dat"; match
(base,/[a-z]+/); seq=substr(base, RSTART,RLENGH)}
{AS[$1]=$2} END{out=""; for(i=0;i<RLENGTH;i+=3)
{triple=substr(seq,i+1,3); out=out AS[triple]};
print out}' code.dat ↵
```

The main work is done by the END section within a loop statement. The base sequence is split into triples by means of an index variable i, which defines the end of the foregoing triple. The function substr locates the triple from the string seq starting at character position i+1, spawning three characters. Next, the corresponding amino acid is retrieved from the array AS with the help of the just calculated variable triple. Next, we connect it with the string out, which holds the amino acids, detected so far. To get the desired result, we make use of the for-loop-statement, which starts after an initialization out="", where an empty output string out is defined. The loop first assigns the value zero to its loop-variable i. The next controlling statement ensures a definite number of loop cycles, given by a relational expression. The expression "i+=3" is evaluated as the last one after each set of the loop statements and increments the controlling variable i by three. So, the loop will be exited, if the value of i becomes equal to the value of RLENGTH. The next pair of braces holds the so-called loop statements, comprising the definition of the variable `triple` and the concatenation of the strings `out` and the amino acid name `AS[triple]` with the help of a blank as the appropriate operator. Note, that in case of more than one statement inside a loop, all the statements have to be placed into braces. As a last step, the whole amino acid sequence is printed out.

Of course, the last `gawk` command exhibits a very complex structure, so typing it on the command line may be very hard especially in the case of spelling mistakes. To avoid such problems, `gawk` is able to interpret a script containing the pattern-action statements. To create such a command script, start an editor and enter the program between the quotes of the last `gawk` command. Take into consideration that the first line of this script must have the following form, if we assume `gawk` resides in the directory `/usr/bin`:

```
#! /usr/bin/gawk -f
```

The shell will start the command `gawk` with the option `-f` and the script name as the next argument. Additional arguments on the command line will be made available to `gawk` in the usual manner.

Assume, the script will be named `sequence`, its contents should look like this:

```
#! /usr/bin/gawk -f

BEGIN{getline base < "base.dat"; match (base,/[a-z]+/);
            seq=substr(base,RSTART,RLENGTH)}
{AS[$1]=$2}
END{out="";
for(i=0;i<RLENGTH;i+=3){triple=substr(seq,i+1,3);
     out=out AS[triple]};print out}
```

To start this script, the user must have read and execute permissions. Thus, use the command `chmod` to add the execute right to the existing read right:

```
> chmod u+x sequence ↵
```

Now, you are able to start the script by typing its name, provided it resides in the actual directory:

```
> sequence code.dat ↵
```

Thus, the shell will execute the following command:

```
/usr/bin/gawk -f sequence code.dat
```

and the output should look like the one you get by typing the whole `gawk` program on the command line.

To get the script more flexible, one would replace the file name base.dat by a variable, let's say "`in`", so the `getline` action near the pattern "`BEGIN`" would read:

```
BEGIN{getline base < in; ....
```

The appropriate command line now looks like that:

```
> sequence -v in=base.dat code.dat ↵
```

the option `-v` tells `gawk` that the next argument is the variable assignment `x=base.dat`, resulting in a replacement of all instances `x` within the script by its value `base.dat`.

Appendix

Summary of Important Internet Resources

Software for Simulation and Other Calculations

Name	Source and short description
Gromacs	http://www.gromacs.org "GROMACS is a versatile package to perform molecular dynamics, i.e. simulate the Newtonian equations of motion for systems with hundreds to millions of particles" (www.gromacs.org) (Scott et al. 1999, van der Spoel et al. 2005)
PRODRG	http://davapc1.bioch.dundee.ac.uk/prodrg/ "will take a description of a small molecule and from it generate a variety of topologies for use with GROMACS . . . ." (http://davapc1.bioch.dundee.ac.uk/prodrg/) (Schuettelkopf and van Aalten 2004)
NAMD	http://www.ks.uiuc.edu/Research/namd/ A parallel MD code for high performance simulations of large systems
TINKER	http://dasher.wustl.edu/ffe Software tools for molecular design
Clustal	http://www.clustal.org Software for multiple sequence alignment
Whatif	http://swift.cmbi.ru.nl/whatif/ A versatile molecular modelling package (Vriend 1990)
PROCHECK	http://www.ebi.ac.uk/thornton-srv/software/PROCHECK Checks stereochemical quality of a protein
PSIPRED	http://bioinf.cs.ucl.ac.uk/psipred/ Protein structure prediction server
I-TASSER	http://zhanglab.ccmb.med.umich.edu/I-TASSER Protein structure and function predictions

A. Strasser, H.-J. Wittmann, *Modelling of GPCRs*,
DOI 10.1007/978-94-007-4596-4, © Springer Science+Business Media Dordrecht 2013

Software for Visualisation

Name	Source and short description
Chimera	http://www.cgl.ucsf.edu/chimera/ A software for visualization of molecular structures
vmd	http://www.ks.uiuc.edu/Research/vmd/ A molecular visualization program
Rasmol	http://rasmol.org A molecular visualization software
xmgrace	http://plasma-gate.weizmann.ac.il/Grace/ 2D data visualisation

Databases

Name	Source and short description
GPCRDB	http://www.gpcr.org/7tm Information system for G protein-coupled receptors
PDB	http://www.pdb.org Archive, containing information about experimentally determind structures of proteins for example
Expasy	http://www.expasy.org "Provides access to scientific databases and software tools." (http://www.expasy.org)
Drug Bank	http://www.drugbank.ca A drug data and drug target database
GPCR network	http://cmpd.scripps.edu Platform of the GPCR community
gpDB	http://bioinformatics2.biol.uoa.gr/gpDB/ A database of GPCRs, G-Proteins, Effectors and their interactions (Elefsinioti et al. 2004;Theodoropoulou et al. 2008)
GPCR-OKB	http://data.gpcr-okb.org/gpcr-okb A database about GPCR oligomerization (Skrabanek et al. 2007; Khelashvili et al. 2010)
IUPHAR	http://www.iuphar-db.org Database on receptor nomenclature and drug classification

Sources with Regard to Lipids

URL
http://lipidbook.bioch.ox.ac.uk
http://moose.bio.ucalgary.ca/index.php?page=Structures_and_Topologies
http://www.lrz-muenchen.de/~heller/membrane/membrane.html
http://www.scmbb.ulb.ac.be/Users/lensink/lipid/

Natural Amino Acids

				pI	*pK$_s$*	*code*
Alanine	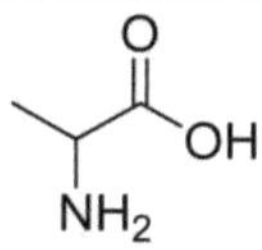	Ala	A	6.0	2.4 9.9	GCU GCC GCA GCG
Arginine	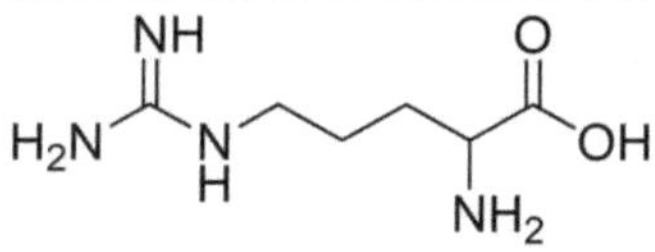	Arg	R	11.2	1.8 9.0 13.2	CGU CGC CGA CGG AGA AGG
Asparagine	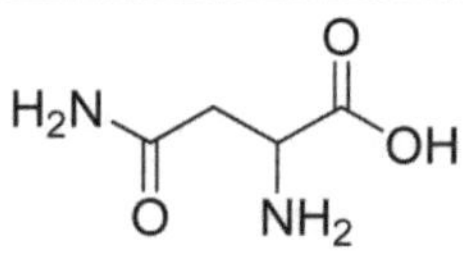	Asn	N	5.4	2.0 8.8	AAU AAC
Aspartic acid	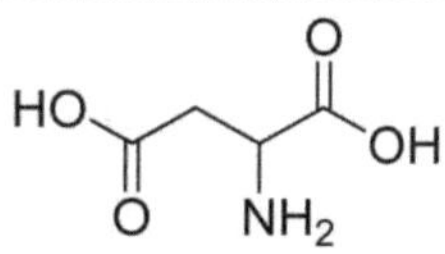	Asp	D	2.8	2.0 3.9 10.0	GAU GAC
Cysteine	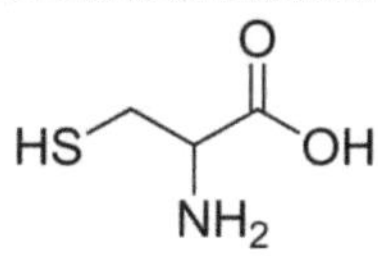	Cys	C	5.0	1.9 10.3	UGU UGC
Glutamine	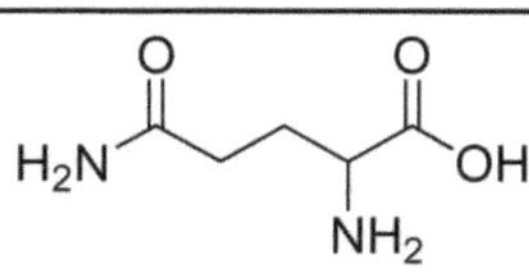	Gln	Q	5.7	2.2 9.1	CAA CAG
Glutamine acid	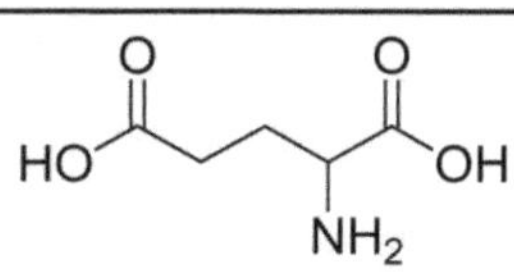	Glu	E	3.2	2.1 4.3 10.0	GAA GAG
Glycine	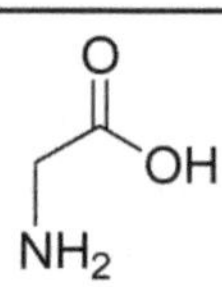	Gly	G	6.0	2.4 9.8	GGU GGC GGA GGG

				pI	pK_s	*code*
Histidine	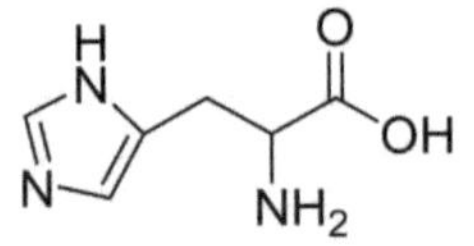	His	H	7.5	1.8 6.1 9.2	CAU CAC
Isoleucine	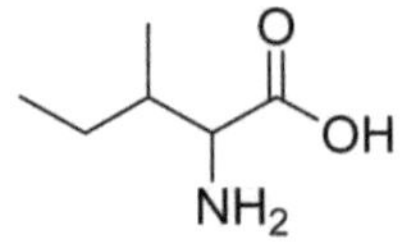	Ile	I	5.9	2.3 9.8	AUU AUC AUA
Leucine	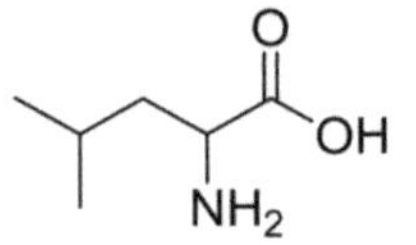	Leu	L	6.0	2.3 9.7	UUA UUG CUU CUC CUA CUG
Lysine	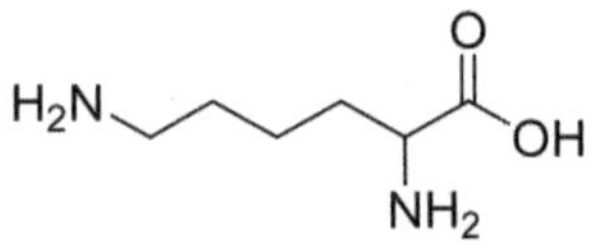	Lys	K	9.6	2.2 9.2 10.8	AAA AAG
Methionine	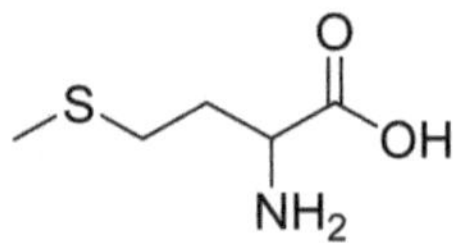	Met	M	5.7	2.2 9.3	AUG
Phenylalanine	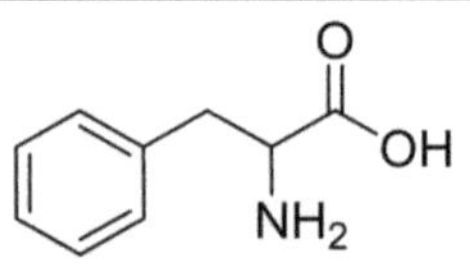	Phe	F	5.5	2.6 9.2	UUU UUC
Proline	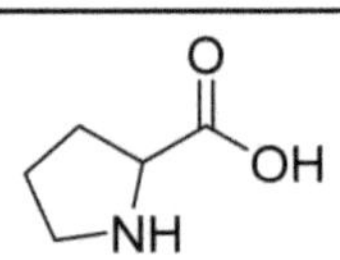	Pro	P	6.3	2.0 10.6	CCU CCC CCA CCG
Serine	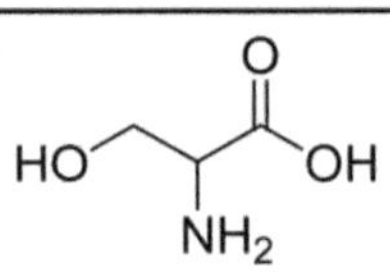	Ser	S	5.7	2.2 9.4	UCU UCC UCA UCG AGU AGC

				pI	pK_s	*code*
Threonine	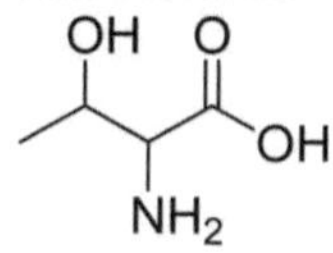	Thr	T	5.6	2.1 9.1	ACU ACC ACA ACG
Tryptophane	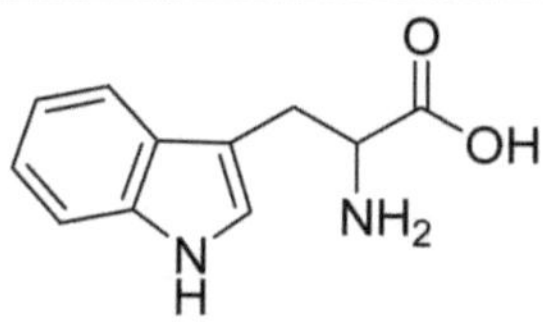	Trp	W	5.9	2.4 9.4	UGG
Tyrosine	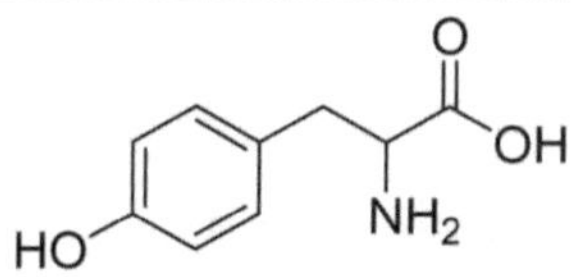	Tyr	Y	5.7	2.2 10.1	UAU UAC
Valine	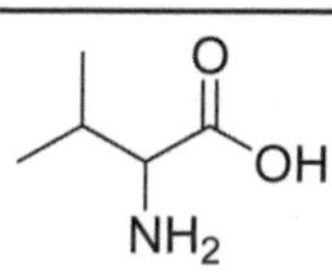	Val	V	6.0	2.3 9.7	GUU GUC GUA GUG

GPCR Families (Source: http://www.gpcr.org/7tm)

- Class A, rhodopsin like
  - Amine
    - Muscarinic acetylcholine
    - Adrenoceptors
    - Dopamine
    - Histamine
    - Serotonin
    - Octopamine
    - Trace amine
  - Peptide
    - Angiotensin
    - Bombesin
    - Bradykinin
    - C5a anaphylatoxin
    - Fmet-leu-phe
    - APJ like

  - Interleukin-8
  - Chemokine
  - Cholecystokinin
  - Endothelin
  - Melanocortin
  - Duffy antigen
  - Prolactin-releasing peptide (GPR10)
  - Neuropeptide Y
  - Neurotensin
  - Opioid
  - Somatostatin
  - Tachykinin
  - Vasopressin-like
  - Galanin like
  - Proteinase-activated like
  - Orexin & neuropeptides FF, QRFP
  - Urotensin II
  - Adrenomedullin (G10D)
  - GPR37/endothelin B-like
  - Chemokine receptor-like
  - Neuromedin U like
  - Somatostatin- and angiogenin-like peptide
  - Allatostatin C/drostatin C
  - Melanin-concentrating hormone receptors
  - Prokineticin receptors
  - Sulfakinin/CCKLR
  - Other peptide receptors
- Hormone protein
  - Follicle stimulating hormone
  - Lutropin-choriogonadotropic hormone
  - Thyrotropin
  - Gonadotropin
- (Rhod)opsin
- Olfactory
- Prostanoid
  - Prostaglandin
  - Prostacyclin
  - Thromboxane
- Nucleotide-like
  - Adenosine
  - Purinoceptors

  - Cannabinoid
  - Platelet activating factor
  - Gonadotropin-releasing hormone
    - Gonadotropin-releasing hormone
    - Adipokinetic hormone like
    - Corazonin
    - Gonadotropin-releasing hormone (other)
  - Thyrotropin-releasing hormone & Secretagogue
    - Thyrotropin-releasing hormone
    - Growth hormone secretagogue
    - Growth hormone secretagogue like
    - Ecdysis-triggering hormone (ETHR)
  - Melatonin
  - Viral
  - Lysosphingolipid & LPA (EDG)
  - Leukotriene B4 receptor
  - Orphan/Other
    - Putative neurotransmitters
    - SREB
    - Mas proto-oncogene & Mas-related (MRGs)
    - RDC1
    - EBV-induced
    - ORPH
    - LGR like (hormone receptors)
    - GPR
    - GPR45 like
    - Cysteinyl leukotriene
    - G-protein coupled bile acid receptor
    - Free fatty acid receptor (GP40, GP41, GP43)
- Class B, secretin like
  - Calcitonin
  - Corticotropin releasing factor
  - Gastric inhibitory peptide
  - Glucagon
  - Growth hormone-releasing hormone
  - Parathyroid hormone
  - PACAP
  - Secretin
  - Vasoactive intestinal polypeptide
  - Diuretic hormone
  - EMR1

- Latrophilin
- Brain-specific angiogenesis inhibitor (BAI)
- Methuselah-like proteins (MTH)
- Cadherin EGF LAG (CELSR)
- Very large G-protein coupled receptor

• Class C, metabotropic glutamate/pheromone)

  - Metabotropic glutamate
  - Calcium-sensing like
  - Putative pheromone receptors
  - GABA-B
  - Orphan GPCR5
  - Orphan GPCR6
  - Bridge of sevenless proteins (BOSS)
  - Taste receptors (T1R)

• Class D, fungal pheromone

  - Fungal pheromone A-factor like (STE2, STE3)
  - Fungal pheromone B like (BAR, BBR, RCB, PRA)
  - Fungal pheromone M- and P-factor

• Class E, cAMP receptors

Listing of Biogenic Amine Receptors

• Muscarinic acetylcholine

  - M_1
  - M_2
  - M_3
  - M_4
  - M_5

• Adrenoceptors

  - Alpha adrenoceptors

    - α_{1a}
    - α_{1b}
    - α_{1d}
    - α_{2a}
    - α_{2b}
    - α_{2c}
    - α_{2d}

  - Beta adrenoceptors

- β_1
- β_2
- β_3
- β_4

– Dopamine receptors

- D_1
- D_2
- D_3
- D_4
- D_5

– Histamine receptors

- H_1
- H_2
- H_3
- H_4

– Serotonin receptors

- 5-HT_{1a}
- 5-HT_{1b}
- 5-HT_{1c}
- 5-HT_{1d}
- 5-HT_{1e}
- 5-HT_{1f}
- 5-HT_{2a}
- 5-HT_{2b}
- 5-HT_{2c}
- 5-HT_4
- 5-HT_{5a}
- 5-HT_{5b}
- 5-HT_6
- 5-HT_7

POPC Parameters

Source: http://moose.bio.ucalgary.ca/index.php?page=Structures_and_Topologies

The file `popc.itp`, available at the internet source mentioned above, is shown below. Please note, that the identifier for the residue `POPC` is changed to `POP` in the related example of Chap. 5. The types of the sites within this file are defined in `lipid.itp`, also available at the same internet source.

```
[ moleculetype ]
; Name   nrexcl
POPC     3

[ atoms ]
;   nr    type   resnr  residu    atom     cgnr          charge
mass
     1     LC3       1     POPC      C1        0          0.4000
15.0350 ; qtot:0.36
     2     LC3       1     POPC      C2        0          0.4000
15.0350 ; qtot:0.72
     3     LC3       1     POPC      C3        0          0.4000
15.0350 ; qtot:1.08
     4     LNL       1     POPC      N4        0         -0.5000
14.0067 ; qtot:0.76
     5     LH2       1     POPC      C5        0          0.3000
14.0270 ; qtot:1.0
     6     LC2       1     POPC      C6        1          0.4000
14.0270 ; qtot:1.0
     7     LOS       1     POPC      O7        1         -0.800
15.9994 ; qtot:0.54
     8     LP        1     POPC      P8        1          1.700
30.9738 ; qtot:2.3
     9     LOM       1     POPC      O9        1         -0.800
15.9994 ; qtot:1.5
    10     LOM       1     POPC     O10        1         -0.800
15.9994 ; qtot:0.7
    11     LOS       1     POPC     O11        1         -0.700
15.9994 ; qtot:0
    12     LC2       1     POPC     C12        2          0.400
14.0270 ; qtot:0.08
    13     LH1       1     POPC     C13        2          0.300
13.0190 ; qtot:0.52
    14     LOS       1     POPC     O14        2         -0.700
15.9994 ; qtot:-0.14
    15      LC       1     POPC     C15        2          0.7000
12.0110 ; qtot:0.56
    16      LO       1     POPC     O16        2         -0.700
15.9994 ; qtot:0.0
    17     LP2       1     POPC     C17        3          0.0
14.0270 ; qtot:
    18     LP2       1     POPC     C18        4               0
14.0270 ; qtot:
    19     LP2       1     POPC     C19        5               0
14.0270 ; qtot:
    20     LP2       1     POPC     C20        6               0
14.0270 ; qtot:
    21     LP2       1     POPC     C21        7               0
14.0270 ; qtot:
    22     LP2       1     POPC     C22        8               0
14.0270 ; qtot:
    23     LP2       1     POPC     C23        9               0
14.0270 ; qtot:
    24     LH1       1     POPC     C24       10               0
13.0190 ; qtot:
    25     LH1       1     POPC     C25       11               0
```

```
13.0190 ; qtot:
    26      LP2        1     POPC      C26        12              0
14.0270 ; qtot:
    27      LP2        1     POPC      C27        13              0
14.0270 ; qtot:
    28      LP2        1     POPC      C28        14              0
14.0270 ; qtot:
    29      LP2        1     POPC      C29        15              0
14.0270 ; qtot:
    30      LP2        1     POPC      C30        16              0
14.0270 ; qtot:
    31      LP2        1     POPC      C31        17              0
14.0270 ; qtot:
    32      LC2        1     POPC      C32        18           0.50
14.0270 ; qtot:
    33      LOS        1     POPC      O33        18          -0.70
15.9994 ; qtot:
    34       LC        1     POPC      C34        18           0.800
12.0110 ; qtot:
    35       LO        1     POPC      O35        18          -0.60
15.9994 ; qtot:
    36      LP2        1     POPC      C36        19              0
14.0270 ; qtot:
    37      LP2        1     POPC      C37        20              0
14.0270 ; qtot:
    38      LP2        1     POPC      C38        21              0
14.0270 ; qtot:
    39      LP2        1     POPC      C39        22              0
14.0270 ; qtot:
    40      LP2        1     POPC      C40        23              0
14.0270 ; qtot:
    41      LP2        1     POPC      C41        24              0
14.0270 ; qtot:
    42      LP2        1     POPC      C42        25              0
14.0270 ; qtot:
    43      LP2        1     POPC      C43        26              0
14.0270 ; qtot:
    44      LP2        1     POPC      C44        27              0
14.0270 ; qtot:
    45      LP2        1     POPC      C45        28              0
14.0270 ; qtot:
    46      LP2        1     POPC      C46        29              0
14.0270 ; qtot:
    47      LP2        1     POPC      C47        30              0
14.0270 ; qtot:
    48      LP2        1     POPC      C48        31              0
14.0270 ; qtot:
    49      LP2        1     POPC      C49        32              0
14.0270 ; qtot:
    50      LP3        1     POPC      C50        33              0
15.0350 ; qtot:
    51      LP2        1     POPC      CA1        34              0
14.0270 ;  tail2
    52      LP3        1     POPC      CA2        35              0
15.0350 ; tail2
```

```
[ bonds ]
;  ai    aj funct
      4       5          1 0.14700E+00 0.37660E+06
      5       6          1 0.15300E+00 0.33470E+06
      6       7          1 0.14300E+00 0.25100E+06
      7       8          1 0.16100E+00 0.25100E+06
      8       9          1 0.14800E+00 0.37660E+06
      8      10          1 0.14800E+00 0.37660E+06
      8      11          1 0.16100E+00 0.25100E+06
     11      12          1 0.14300E+00 0.25100E+06
     12      13          1 0.15300E+00 0.33470E+06
     13      14          1 0.14350E+00 0.25100E+06
     13      32          1 0.15300E+00 0.33470E+06
     14      15          1 0.13600E+00 0.37660E+06
     15      16          1 0.12300E+00 0.50210E+06
     15      17          1 0.15300E+00 0.33470E+06
     17      18          1 0.15300E+00 0.33470E+06
     18      19          1 0.15300E+00 0.33470E+06
     19      20          1 0.15300E+00 0.33470E+06
     20      21          1 0.15300E+00 0.33470E+06
     21      22          1 0.15300E+00 0.33470E+06
     22      23          1 0.15300E+00 0.33470E+06
     23      24          1 0.15300E+00 0.33470E+06
     24      25          1 0.13900E+00 0.41840E+06
     25      26          1 0.15300E+00 0.33470E+06
     26      27          1 0.15300E+00 0.33470E+06
     27      28          1 0.15300E+00 0.33470E+06
     28      29          1 0.15300E+00 0.33470E+06
     29      30          1 0.15300E+00 0.33470E+06
     30      31          1 0.15300E+00 0.33470E+06
     31      51          1 0.15300E+00 0.33470E+06
     51      52          1 0.15300E+00 0.33470E+06
     32      33          1 0.14300E+00 0.25100E+06
     33      34          1 0.13600E+00 0.37660E+06
     34      35          1 0.12300E+00 0.50210E+06
     34      36          1 0.15300E+00 0.33470E+06
     36      37          1 0.15300E+00 0.33470E+06
     37      38          1 0.15300E+00 0.33470E+06
     38      39          1 0.15300E+00 0.33470E+06
     39      40          1 0.15300E+00 0.33470E+06
     40      41          1 0.15300E+00 0.33470E+06
     41      42          1 0.15300E+00 0.33470E+06
     42      43          1 0.15300E+00 0.33470E+06
     43      44          1 0.15300E+00 0.33470E+06
     44      45          1 0.15300E+00 0.33470E+06
     45      46          1 0.15300E+00 0.33470E+06
     46      47          1 0.15300E+00 0.33470E+06
     47      48          1 0.15300E+00  0.33470E+06
     48      49          1 0.15300E+00  0.33470E+06
     49      50          1 0.15300E+00  0.33470E+06
      1       4          1 0.14700E+00  0.37450E+06
      2       4          1 0.14700E+00  0.37450E+06
      3       4          1 0.14700E+00  0.37450E+06
```

```
[ pairs ]
;  ai    aj funct
    1     6     1
    2     6     1
    3     6     1
    4     7     1
    5     8     1
    6     9     1
    6    10     1
    6    11     1
    7    12     1
    8    13     1
    9    12     1
   10    12     1
   11    14     1
   11    32     1
   12    15     1
   12    33     1
   13    16     1
   13    17     1
   13    34     1
   14    18     1
   14    33     1
   15    19     1
   15    32     1
   16    18     1
   22    25     1     ; pair around double bond
   24    27     1     ; pair around double bond
   32    35     1
   32    36     1
   33    37     1
   34    38     1
   35    37     1

[ angles ]
;  ai    aj    ak funct
       4        5        6        1 0.10950E+03 0.46020E+03
       5        6        7        1 0.10950E+03 0.46020E+03
       6        7        8        1 0.12000E+03 0.39750E+03
       7        8        9        1 0.10960E+03 0.39750E+03
       7        8       10        1 0.10960E+03 0.39750E+03
       7        8       11        1 0.10300E+03 0.39750E+03
       8       11       12        1 0.12000E+03 0.39750E+03
       9        8       10        1 0.12000E+03 0.58580E+03
       9        8       11        1 0.10960E+03 0.39750E+03
      10        8       11        1 0.10960E+03 0.39750E+03
      11       12       13        1 0.11100E+03 0.46020E+03
      12       13       14        1 0.10950E+03 0.46020E+03
      12       13       32        1 0.10950E+03 0.46020E+03
      13       14       15        1 0.12000E+03 0.41840E+03
      13       32       33        1 0.11100E+03 0.46020E+03
      14       13       32        1 0.10950E+03 0.46020E+03
      14       15       16        1 0.12400E+03 0.50210E+03
      14       15       17        1 0.11500E+03 0.50210E+03
      15       17       18        1 0.11100E+03 0.46020E+03
      16       15       17        1 0.12100E+03 0.50210E+03
      17       18       19        1 0.11100E+03 0.46020E+03
      18       19       20        1 0.11100E+03 0.46020E+03
      19       20       21        1 0.11100E+03 0.46020E+03
      20       21       22        1 0.11100E+03 0.46020E+03
      21       22       23        1 0.11100E+03 0.46020E+03
      22       23       24        1 0.11100E+03 0.46020E+03
```

```
   23          24       25         1 120.000      502.080   ; cis thingies
   24          25       26         1 120.000      502.080   ; cis thingies
   25          26       27         1 0.11100E+03 0.46020E+03
   26          27       28         1 0.11100E+03 0.46020E+03
   27          28       29         1 0.11100E+03 0.46020E+03
   28          29       30         1 0.11100E+03 0.46020E+03
   29          30       31         1 0.11100E+03 0.46020E+03
   30          31       51         1 0.11100E+03 0.46020E+03
   31          51       52         1 0.11100E+03 0.46020E+03
   32          33       34         1 0.12000E+03 0.41840E+03
   33          34       35         1 0.12400E+03 0.50210E+03
   33          34       36         1 0.11500E+03 0.50210E+03
   34          36       37         1 0.11100E+03 0.46020E+03
   35          34       36         1 0.12100E+03 0.50210E+03
   36          37       38         1 0.11100E+03 0.46020E+03
   37          38       39         1 0.11100E+03 0.46020E+03
   38          39       40         1 0.11100E+03 0.46020E+03
   39          40       41         1 0.11100E+03 0.46020E+03
   40          41       42         1 0.11100E+03 0.46020E+03
   41          42       43         1 0.11100E+03 0.46020E+03
   42          43       44         1 0.11100E+03 0.46020E+03
   43          44       45         1 0.11100E+03 0.46020E+03
   44          45       46         1 0.11100E+03 0.46020E+03
   45          46       47         1 0.11100E+03 0.46020E+03
   46          47       48         1 0.11100E+03 0.46020E+03
   47          48       49         1 0.11100E+03 0.46020E+03
   48          49       50         1 0.11100E+03 0.46020E+03
    1           4        2         1 0.10950E+03 0.33470E+03
    2           4        3         1 0.10950E+03 0.33470E+03
    3           4        1         1 0.10950E+03 0.33470E+03
    1           4        5         1 0.10950E+03 0.37660E+03
    2           4        5         1 0.10950E+03 0.37660E+03
    3           4        5         1 0.10950E+03 0.37660E+03

[ dihedrals ]
;  ai     aj      ak      al  funct    phi0      cp      mult
    1      4       5       6      1     0.0      3.76    3
    4      5       6       7      1     0.0      5.85    3
    5      6       7       8      1     0.0      3.76    3
    6      7       8      11      1     0.0      1.05    3
    6      7       8      11      1     0.0      3.14    2
    7      8      11      12      1     0.0      1.05    3
    7      8      11      12      1     0.0      3.14    2
    8     11      12      13      1     0.0      3.76    3
   11     12      13      14      1     0.0      2.09    2
   11     12      13      32      1     0.0      5.85    3
   11     12      13      32      1     0.0      0.42    2
   12     13      32      33      1     0.0      5.85    3
   12     13      32      33      1     0.0      0.42    2
   12     13      14      15      1     0.0      3.77    3
   13     32      33      34      1     0.0      3.76    3
   13     14      15      17      1     180.0   16.74    2
   14     13      32      33      1     0.0      2.09    2
   14     15      17      18      1     0.0      0.42    6
   15     17      18      19      1     0.0      5.86    3
   17     18      19      20      3
   18     19      20      21      3
   19     20      21      22      3
   20     21      22      23      3
   21     22      23      24      3
   22     23      24      25     10.000          5.858   3
```

```
;  23     24     25     26      1
   24     25     26     27      1 0.000        5.858        3
   25     26     27     28      3
   26     27     28     29      3
   27     28     29     30      3
   28     29     30     31      3
   29     30     31     51      3
   30     31     51     52      3
   13     32     33     34      1    0.0      3.76     3
   32     33     34     36      1    180.0    16.74    2
   33     34     36     37      1    0.0      0.42     6
   34     36     37     38      1    0.0      5.86     3
   36     37     38     39      3
   37     38     39     40      3
   38     39     40     41      3
   39     40     41     42      3
   40     41     42     43      3
   41     42     43     44      3
   42     43     44     45      3
   43     44     45     46      3
   44     45     46     47      3
   45     46     47     48      3
   46     47     48     49      3
   47     48     49     50      3

[ dihedrals ]
;  ai    aj    ak    al funct
      13      14      32      12       2 35.264       0.33470E+03
      15      14      17      16       2 0.00000E+00 0.16740E+03
      34      33      36      35       2 0.00000E+00 0.16740E+03
      23      24      25      26       2 0.000      167.360

#ifdef POSRES_LIPID
#include "lipid_posre.itp"
#endif
```

Important Crystal Structures of GPCRs (Source: http://www.pdb.org)

bovine rhodopsin	
pdb-code	1F88
method	X-ray diffraction
resolution	2.80 Å
molecule	rhodopsin
fragment	no information at protein data bank (http://www.pdb.org)
mutation	no information at protein data bank (http://www.pdb.org)
ligand	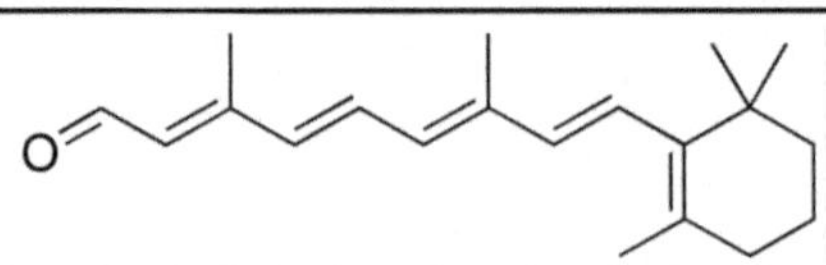
UniProtKB	P02699
literature	Palczewski et al, 2000
pdb-code	1HZX
method	X-ray diffraction
resolution	2.80 Å
molecule	rhodopsin
fragment	no information at protein data bank (http://www.pdb.org)
mutation	no information at protein data bank (http://www.pdb.org)
ligand	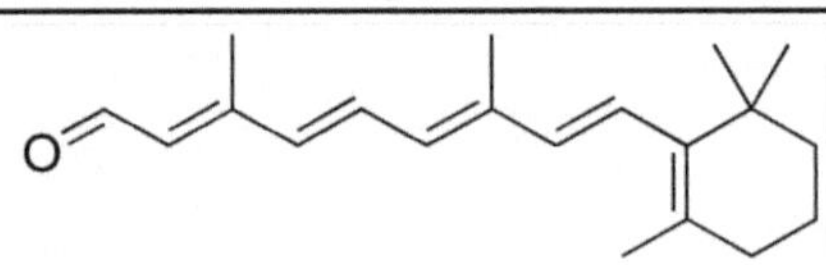
UniProtKB	P02699
literature	Teller et al, 2001

pdb-code	1GZM
method	X-ray diffraction
resolution	2.65 Å
molecule	rhodopsin
fragment	no information at protein data bank (http://www.pdb.org)
mutation	no information at protein data bank (http://www.pdb.org)
ligand	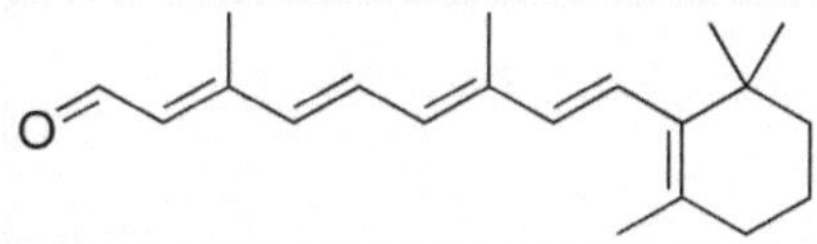
UniProtKB	P02699
literature	Li et al, 2004
pdb-code	3CAP
method	X-rax diffraction
resolution	2.90 Å
molecule	rhodopsin
fragment	no information at protein data bank (http://www.pdb.org)
mutation	no information at protein data bank (http://www.pdb.org)
ligand	this protein is ligand-free rhodopsin, opsin
UniProtKB	P02699
literature	Park et al, 2008
pdb-code	3DQB
method	X-ray diffraction
resolution	3.20 Å
molecule	rhodopsin
fragment	no information at protein data bank (http://www.pdb.org)
mutation	no information at protein data bank (http://www.pdb.org)
ligand	this protein is ligand-free rhodopsin, opsin
UniProtKB	P02699
molecule	11meric peptide from guanine nucleotide-binding protein G(t) sub-unit α_1
fragment	C-terminal domain, residues 340-350
mutation	K341L
UniProtKB	P04695
literature	Scheerer et al, 2008

pdb-code	3PQR
method	X-ray diffraction
resolution	2.85 Å
molecule	rhodopsin
fragment	no information at protein data bank (http://www.pdb.org)
mutation	no information at protein data bank (http://www.pdb.org)
ligand	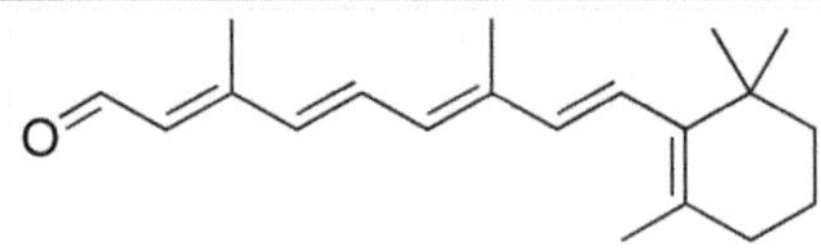
UniProtKB	P02699
molecule	guanine nucleotide-binding protein G(t) subunit α_1
fragment	C-terminal peptide, residues 340-350
mutation	K341L, C347V
UniProtKB	P04695
literature	Choe et al, 2011
pdb-code	3PXO
method	X-ray diffraction
resolution	3.00 Å
molecule	rhodopsin
fragment	no information at protein data bank (http://www.pdb.org)
mutation	no information at protein data bank (http://www.pdb.org)
Ligand	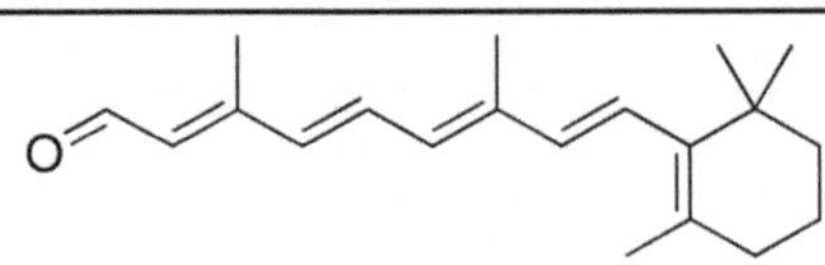
UniProtKB	P02699
literature	Choe et al, 2011

human β_2 adrenergic receptor	
pdb-code	2RH1
method	X-ray diffraction
resolution	2.40 Å
molecule	β_2-adrenergic receptor / T4-lysozyme chimera
fragment	no information at protein data bank (http://www.pdb.org)
mutation	N187E, C54T, C97A
ligand	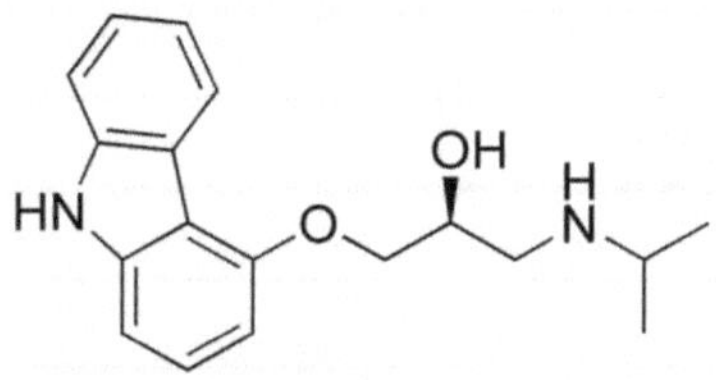
UniProtKB	P07550
literature	Cherezov et al, 2007
pdb-code	2R4R
method	X-ray diffraction
resolution	3.40 Å
molecule	β_2 adrenergic receptor
fragment	no information at protein data bank (http://www.pdb.org)
mutation	no information at protein data bank (http://www.pdb.org)
ligand	without ligand
UniProtKB	P07550
literature	Rasmussen et al, 2007

pdb-code	2R4S
method	X-ray diffraction
resolution	3.40 Å
molecule	β_2 adrenergic receptor
fragment	no information at protein data bank (http://www.pdb.org)
mutation	no information at protein data bank (http://www.pdb.org)
ligand	without ligand
UniProtKB	P07550
literature	Rasmussen et al, 2007

pdb-code	3D4S
method	X-ray diffraction
resolution	2.80 Å
molecule	β_2-adrenergic receptor / T4-lysozyme chimera
fragment	no information at protein data bank (http://www.pdb.org)
mutation	E122W, N187E, C1054T, C1097A
ligand	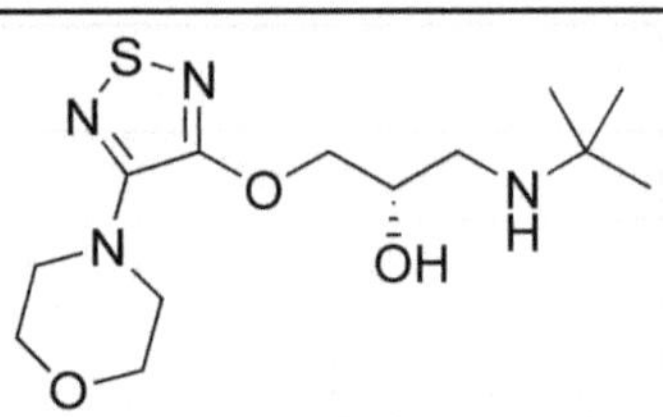 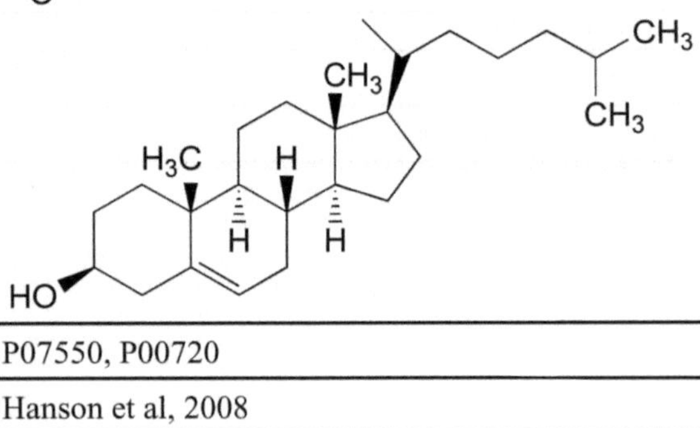
UniProtKB	P07550, P00720
literature	Hanson et al, 2008

pdb-code	3NYA
method	X-ray diffraction
resolution	3.16 Å
molecule	β_2 adrenergic receptor / lysozyme
fragment	chimeric protein of β_2 adrenoceptor 1-230, lysozyme 2-161, β_2-adrenergic receptor 263-348
mutation	E122W, N187E, C1054T, C1097A
ligand	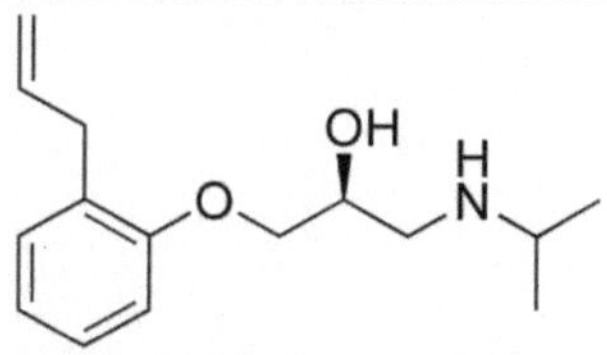
UniProtKB	P07550, P00720
literature	Wacker et al, 2010

pdb-code	3NY8
method	X-ray diffraction
resolution	2.84 Å
molecule	β_2-adrenergic receptor / lysozyme
fragment	chimeric protein of β_2 adrenoreceptor 1-230, lysozyme 2-161, β_2 adrenergic receptor 263-348
mutation	E122W, N187E, C1054T, C1097A
ligand	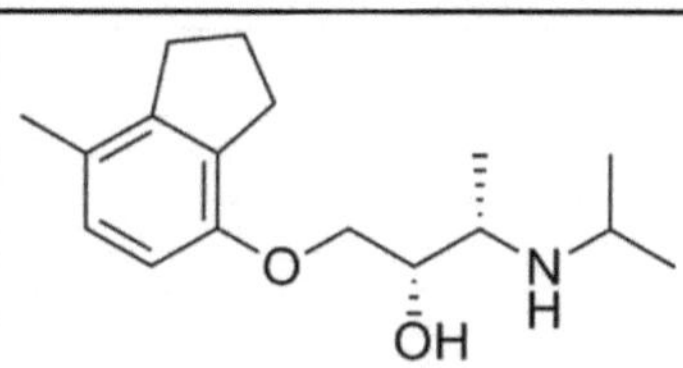
UniProtKB	P07550, P00720
literature	Wacker et al, 2010

pdb-code	3NY9
method	X-ray diffraction
resolution	2.84 Å
molecule	β_2 adrenergic receptor / lysozyme
fragment	chimeric protein of β_2 adrenoceptor 1-230, lysozyme 2-161, β_2-adrenergic receptor 263-348
mutation	E122W, N187E, C1054T, C1097A
ligand	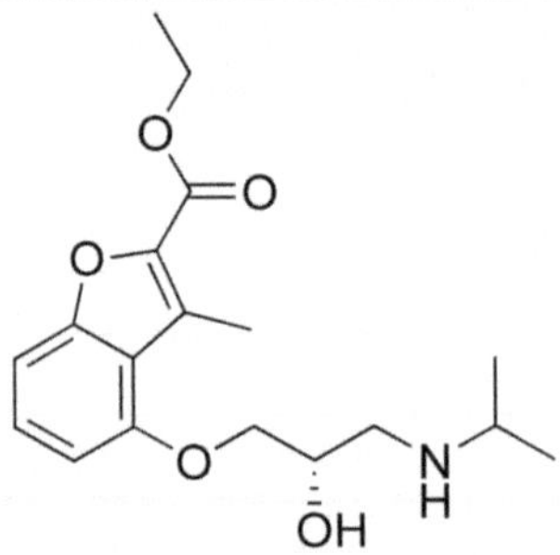
UniProtKB	P07550, P00720
literature	Wacker et al, 2010

pdb-code	3KJ6
method	X-ray diffraction
resolution	3.40 Å
molecule	β_2 adrenergic receptor / Fab light chain / Fab heavy chain
fragment	no information at protein data bank (http://www.pdb.org)
mutation	no information at protein data bank (http://www.pdb.org)
ligand	without ligand
UniProtKB	P07550
literature	Bokoch et al, 2010

pdb-code	3P0G
method	X-ray diffraction
resolution	3.50 Å
molecule	β_2 adrenergic receptor / T4-lysozyme chimera
fragment	P07550 residues 1-230, 263-365; P00720 residues 2-161
mutation	N187E
ligand	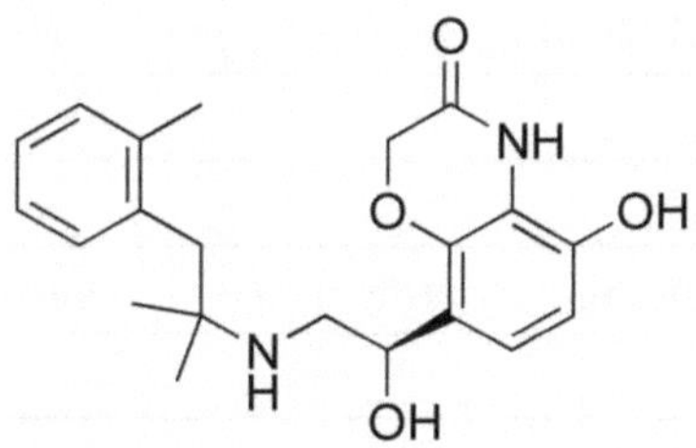
UniProtKB	P07550, P00720
literature	Rasmussen et al, 2011
pdb-code	3PDS
method	X-ray diffraction
resolution	3.50 Å
molecule	fusion protein β_2 adrenergic receptor / lysozyme
fragment	no information at protein data bank (http://www.pdb.org)
mutation	H93C, N187E, C265A
ligand	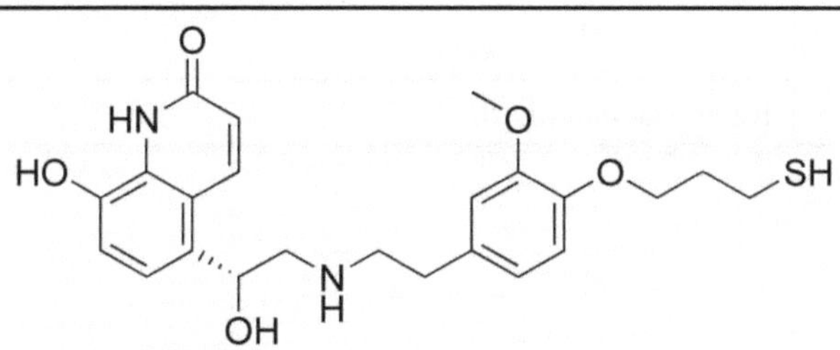
UniProtKB	P07550, P00720
literature	Rosenbaum et al, 2011

pdb-code	3SN6
method	X-ray diffraction
resolution	3.20 Å
molecule	guanine nucleotide-binding protein G(s) subunit α isoforms short
mutation	G72S
UniProtKB	P04896
molecule	guanine nucleotide-binding protein G(i)/G(s)/G(t) subunit β_1
mutation	M1Q
UniProtKB	P54311
molecule	guanine nucleotide-binding protein G(i)/G(s)/G(o) subunit γ_2
UniProtKB	P63212
molecule	lysozyme, β_2-adrenergic receptor
mutation	C54T, C97A, M96T, M98T, N187E
UniProtKB	P07550, P00720
molecule	camelid antibody VHH fragment
ligand	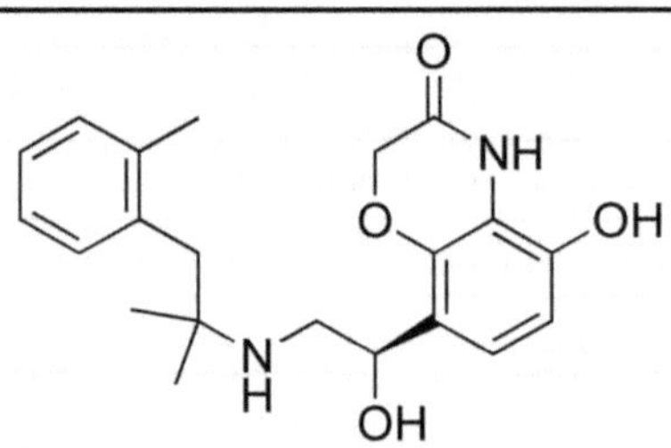
literature	Rasmussen et al, 2011

turkey β_1 adrenergic receptor	
pdb-code	2VT4
method	X-ray diffraction
resolution	2.70 Å
molecule	β_1 adrenergic receptor
fragment	residues 33-243, 272-276, 279-367
mutation	yes
ligand	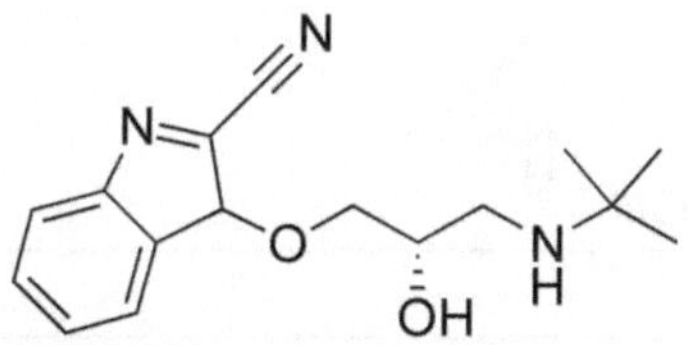
UniProtKB	P07700
literature	Warne et al, 2008
pdb-code	2YCW
method	X-ray diffraction
resolution	3.00 Å
molecule	β_1 adrenergic receptor
fragment	residues 33-243, 272-276, 279-367
mutation	yes
ligand	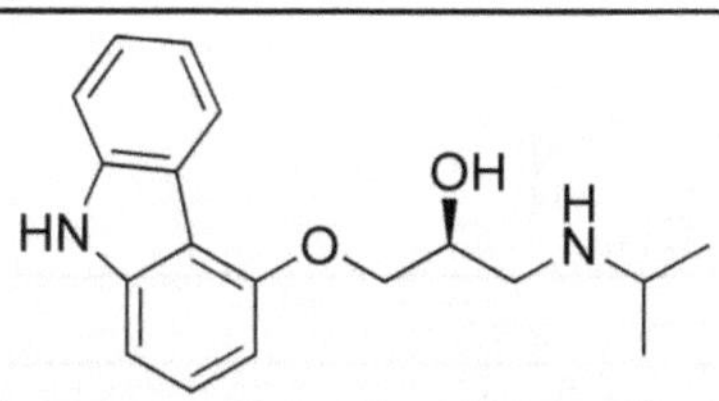
UniProtKB	P07700
literature	Moukhametzianov et al, 2011

pdb-code	2YCX
method	X-ray diffraction
resolution	3.25 Å
molecule	β_1 adrenergic receptor
fragment	residues 33-243, 272-276, 279-367
mutation	yes
ligand	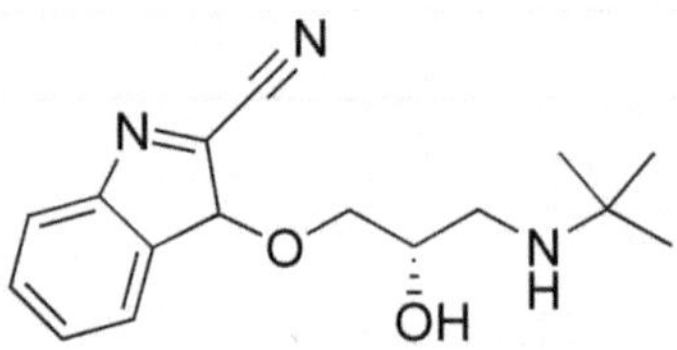
UniProtKB	P07700
literature	Moukhametzianov et al, 2011
pdb-code	2YCY
method	X-ray diffraction
resolution	3.15 Å
molecule	β_1 adrenergic receptor
fragment	residues 33-243, 272-276, 279-367
mutation	yes
ligand	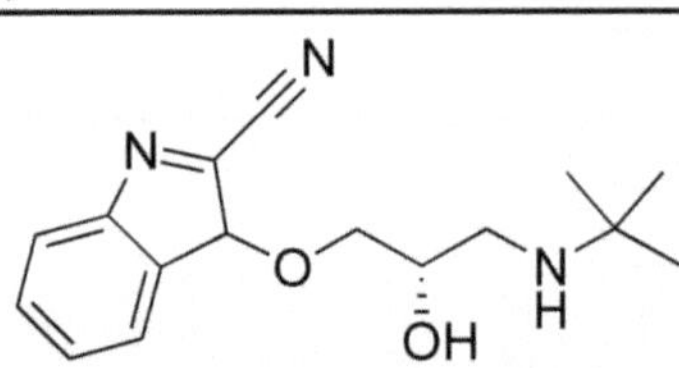
UniProtKB	P07700
literature	Moukhametzianov et al, 2011

pdb-code	2YCZ
method	X-ray diffraction
resolution	3.65 Å
molecule	β_1 adrenergic receptor
fragment	residues 33-243, 272-276, 279-367
mutation	yes
ligand	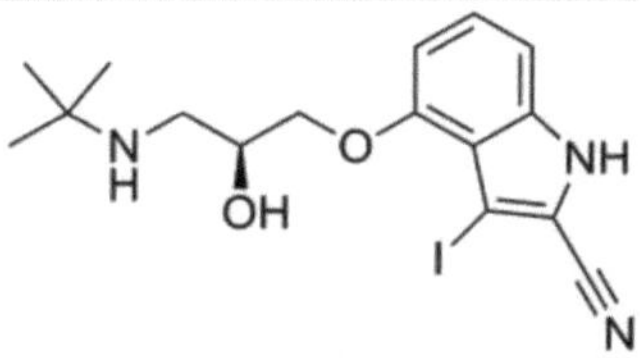
UniProtKB	P07700
literature	Moukhametzianov et al, 2011

pdb-code	2Y00
method	X-ray diffraction
resolution	2.50 Å
molecule	β_1 adrenergic receptor
fragment	residues 33-368
mutation	yes
ligand	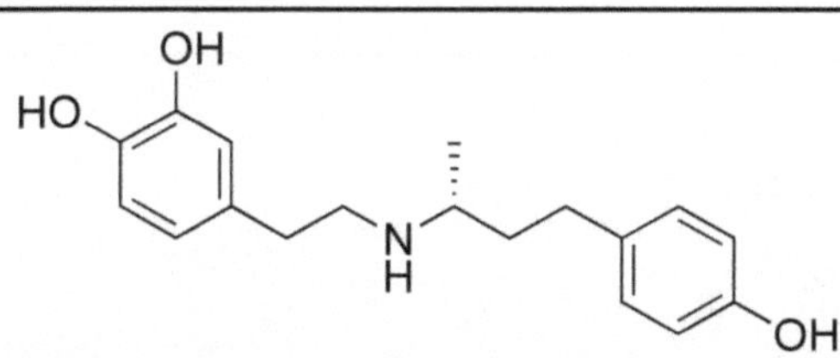
UniProtKB	P07700
literature	Warne et al, 2011

pdb-code	2Y01
method	X-ray diffraction
resolution	2.60 Å
molecule	β_1 adrenergic receptor
fragment	residues 33-368
mutation	yes
ligand	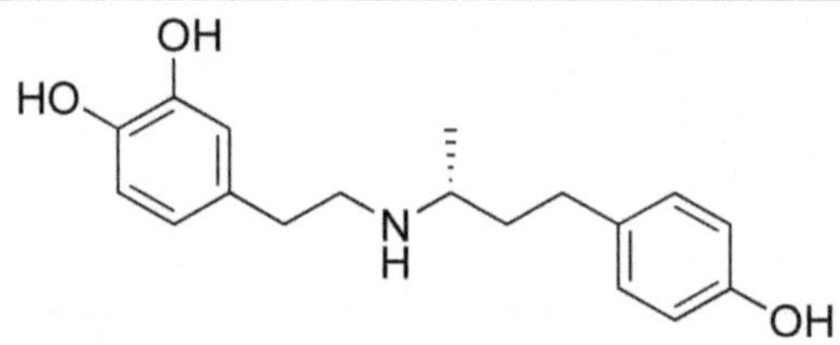
UniProtKB	P07700
literature	Warne et al, 2011
pdb-code	2Y02
method	X-ray diffraction
resolution	2.60 Å
molecule	β_1 adrenergic receptor
fragment	residues 33-368
mutation	yes
ligand	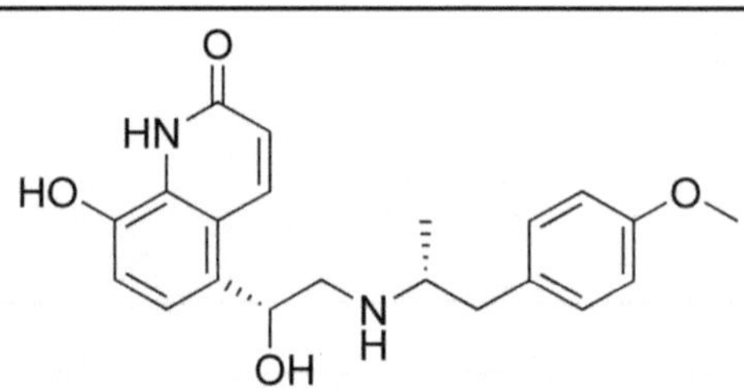
UniProtKB	P07700
literature	Warne et al, 2011

pdb-code	2Y03
method	X-ray diffraction
resolution	2.85 Å
molecule	β_1 adrenergic receptor
fragment	residues 33-368
mutation	yes
ligand	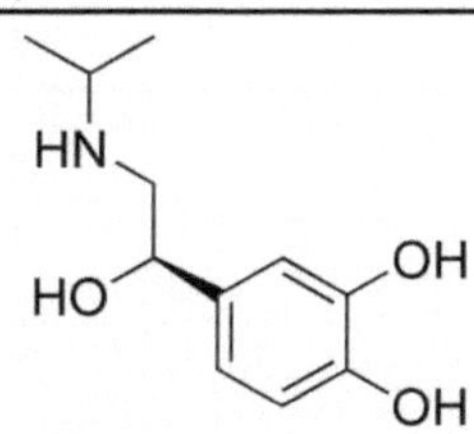
UniProtKB	P07700
literature	Warne et al, 2011
pdb-code	2Y04
method	X-ray diffraction
resolution	3.05 Å
molecule	β_1 adrenergic receptor
fragment	residues 33-368
mutation	yes
ligand	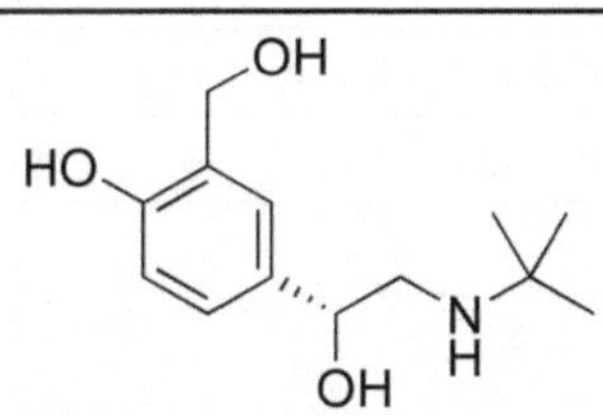
UniProtKB	P07700
literature	Warne et al, 2011

human D_3 receptor	
pdb-code	3PBL
method	X-ray diffraction
resolution	2.89 Å
molecule	D_3 dopamine receptor / lysozyme chimera
fragment	no information at protein data bank (http://www.pdb.org)
mutation	L119W, C1054T, C1097A
ligand	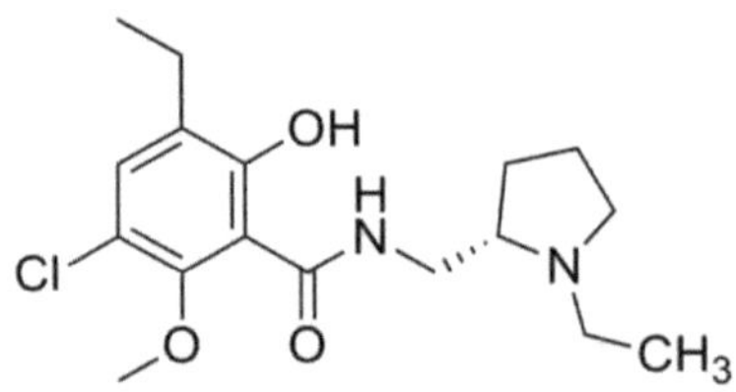
UniProtKB	P00720,P35462
literature	Chien et al, 2010

human H_1 receptor	
pdb-code	3RZE
method	X-ray diffraction
resolution	3.10 Å
molecule	histamine H_1 receptor / lysozyme chimera
fragment	no information at protein data bank (http://www.pdb.org)
mutation	no information at protein data bank (http://www.pdb.org)
ligand	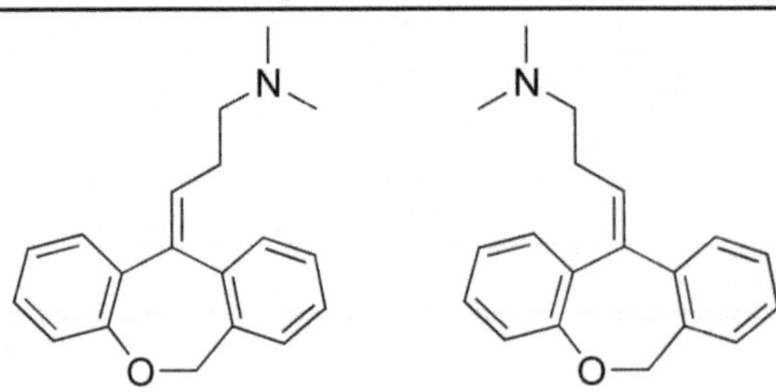
UniProtKB	P35367, P00720
literature	Shimamura et al, 2011

human $CXCR_4$ receptor	
pdb-code	3ODU
method	X-ray diffraction
resolution	2.50 Å
molecule	C-X-C chemokine receptor type 4 / lysozyme chimera
fragment	CXCR4 residues 2-229, lysozyme residues 1002-1161, CXCR4 residues 230-319
mutation	L125W, C1054T, C1097T
ligand	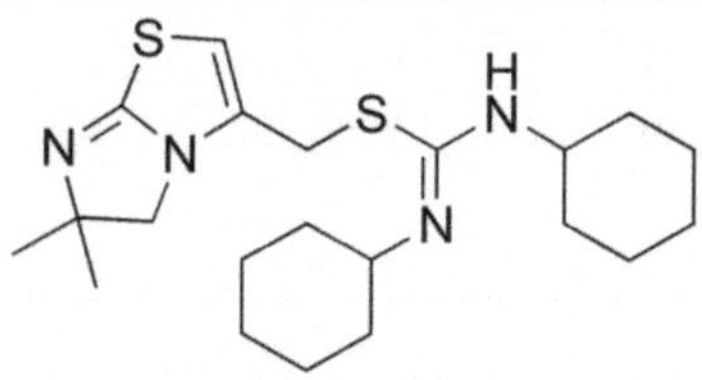
UniProtKB	P61073, P00720
literature	Wu et al, 2010
pdb-code	3OE6
method	X-ray diffraction
resolution	3.20 Å
molecule	C-X-C chemokine receptor type 4 / lysozyme chimera
fragment	CXCR4 residues 2-228, lysozyme residues 1002-1161, CXCR4 residues 231-325
mutation	L125W, C1054T, C1097T
ligand	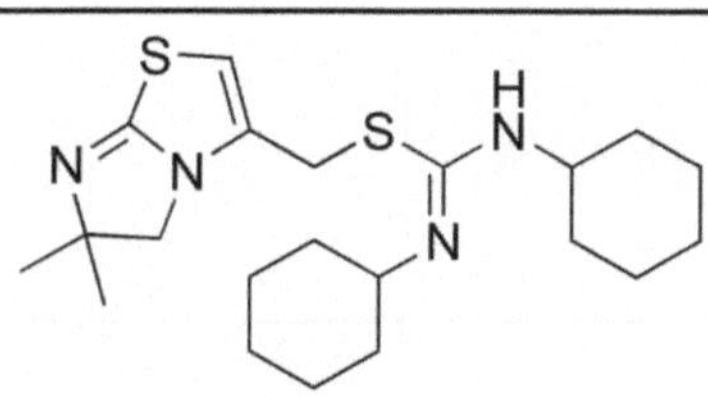
UniProtKB	P61073, P00720
literature	Wu et al, 2010

pdb-code	3OE8
method	X-ray diffraction
resolution	3.10 Å
molecule	C-X-C chemokine receptor type 4 / lysozyme chimera
fragment	CXCR4 residues 2-229, lysozyme residues 1002-1161, CXCR4 residues 231-319
mutation	L125W, C1054T, C1097T
ligand	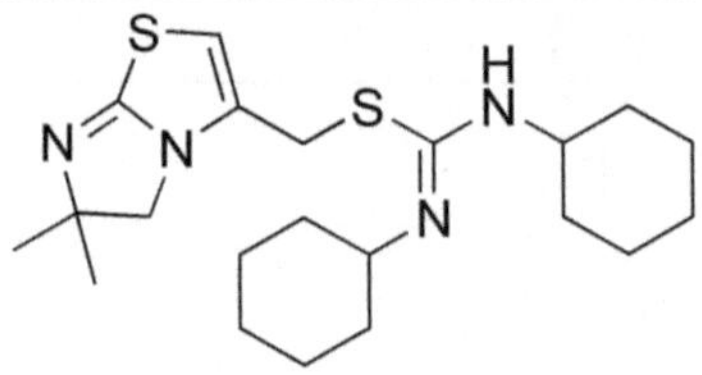
UniProtKB	P61073, P00720
literature	Wu et al, 2010

pdb-code	3OE9
method	X-ray diffraction
resolution	3.10 Å
molecule	C-X-C chemokine receptor type 4 / lysozyme chimera
fragment	CXCR4 residues 2-228, lysozyme residues 1002-1161, CXCR4 residues 231-319
mutation	L125W, T240P
ligand	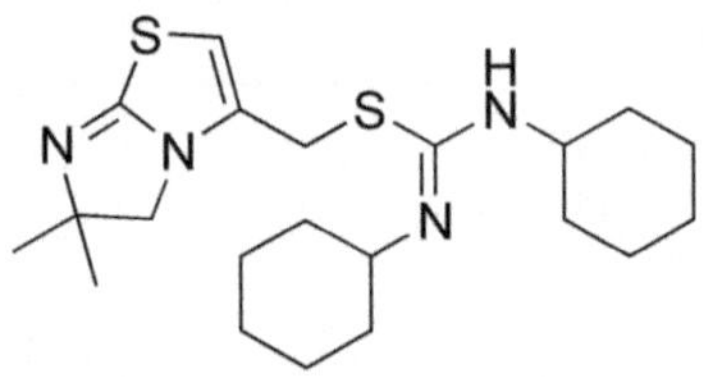
UniProtKB	P61073, P00720
literature	Wu et al, 2010

pdb-code	3OE0
method	X-ray diffraction
resolution	2.90 Å
molecule	C-X-C chemokine receptor type 4 / lysozyme chimera
fragment	CXCR4 residues 2-228, Lysozyme residues 1002-1161, CXCR4 residues 231-319
mutation	L125W, T240P, C1054T, C1097T
ligand	Cyclic peptide CVX15
UniProtKB	P61073, P00720
literature	Wu et al, 2010

human Adenosine A_{2A} receptor	
pdb-code	3EML
method	X-ray diffraction
resolution	2.60 Å
molecule	human adenosine A_{2A} receptor / T4 lysozyme chimera
fragment	no information at protein data bank (http://www.pdb.org)
mutation	no information at protein data bank (http://www.pdb.org)
ligand	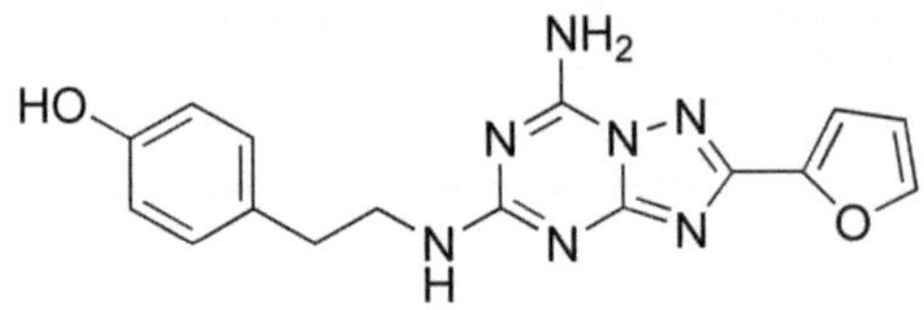
UniProtKB	P29274
literature	Jaakola et al, 2008
pdb-code	2YDO
method	X-ray diffraction
resolution	3.00 Å
molecule	adenosine receptor A_{2A}
fragment	residues 1-317
mutation	yes
ligand	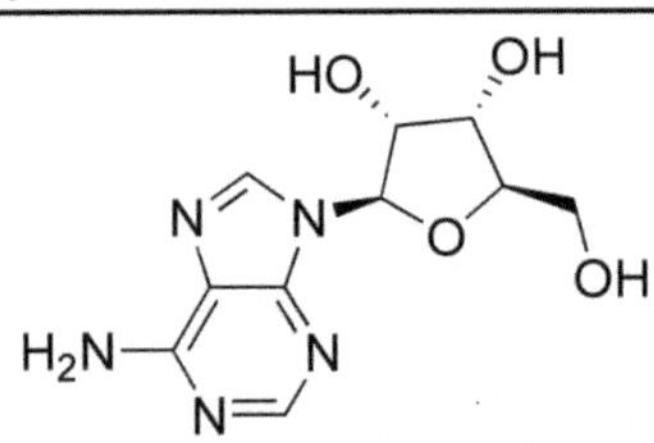
UniProtKB	P29274
literature	Lebon et al, 2011

pdb-code	2YDV
method	X-ray diffraction
resolution	2.60 Å
molecule	adenosine receptor A_{2A}
fragment	residues 1-317
mutation	yes
ligand	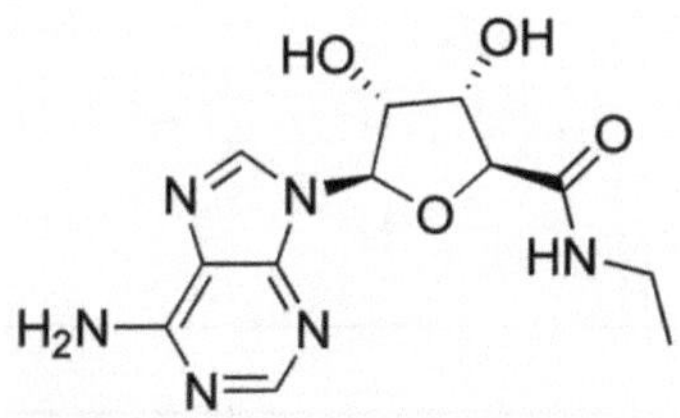
UniProtKB	P29274
literature	Lebon et al, 2011

pdb-code	3QAK
method	X-ray diffraction
resolution	2.71 Å
molecule	adenosine receptor A_{2A} / lysozyme chimera
fragment	no information at protein data bank (http://www.pdb.org)
mutation	no information at protein data bank (http://www.pdb.org)
ligand	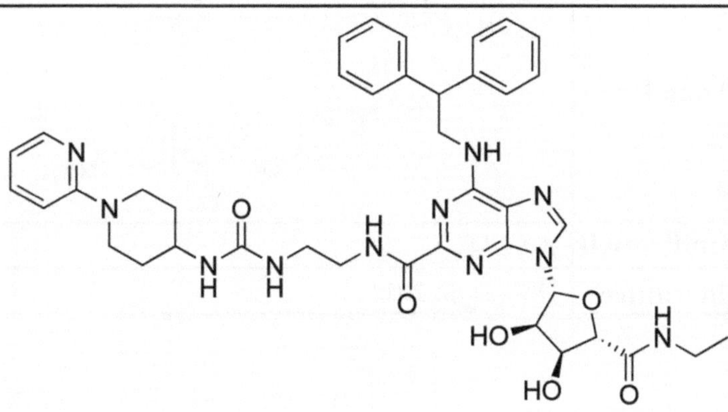
UniProtKB	P29274
literature	Xu et al, 2011

pdb-code	3PWH
method	X-ray diffraction
resolution	3.30 Å
molecule	adenosine receptor A_{2A}
fragment	residues 1-317
mutation	A54L, T88A, K122A, V239A, R107A, L202A,L235A, S277A
ligand	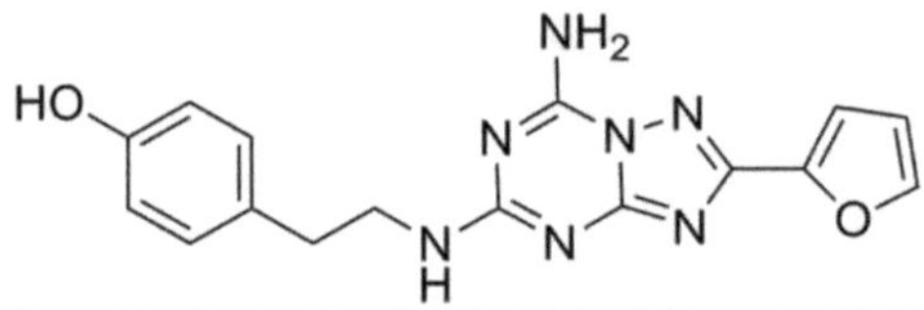
UniProtKB	P29274
literature	Dore et al, 2011
pdb-code	3REY
method	X-ray diffraction
resolution	3.31 Å
molecule	adenosine receptor A_{2A} / lysozyme chimera
fragment	residues 1-317
mutation	A54L, T88A, R107A, K122A, L202A, L235A, V239A, S277A
ligand	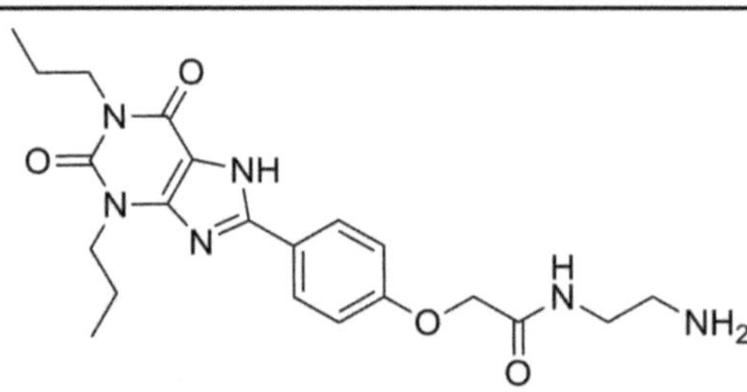
UniProtKB	P29274
literature	Dore et al, 2011

pdb-code	3RFM
method	X-ray diffraction
resolution	3.60 Å
molecule	adenosine receptor A_{2A}
fragment	residues 1-317
mutation	A54L, T88A, R107A, K122A, L202A, L235A, V239A, S277A
ligand	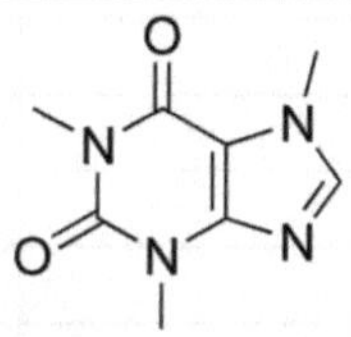
UniProtKB	P29274
literature	Dore et al, 2011

Important Amino Acid Sequences Related to the Crystal Structures of GPCRs (Source: http://www.expasy.org)

bovine (rhod)opsin			
UniProtKB	P02699		
length	348 amino acids		
N-terminus	1-36	**E2-loop**	174-202
TM I	37-63	**TM V**	203-224
C1-loop	64-73	**C3-loop**	225-249
TM II	74-96	**TM VI**	250-274
E1-loop	97-110	**E3-loop**	275-286
TM III	111-133	**TM VII**	287-308
C2-loop	134-152	**C-terminus**	309-348
TM IV	153-173		
sequence			

```
        10         20         30         40         50         60
MNGTEGPNFY VPFSNKTGVV RSPFEAPQYY LAEPWQFSML AAYMFLLIML GFPINFLTLY

        70         80         90        100        110        120
VTVQHKKLRT PLNYILLNLA VADLFMVFGG FTTTLYTSLH GYFVFGPTGC NLEGFFATLG

       130        140        150        160        170        180
GEIALWSLVV LAIERYVVVC KPMSNFRFGE NHAIMGVAFT WVMALACAAP PLVGWSRYIP

       190        200        210        220        230        240
EGMQCSCGID YYTPHEETNN ESFVIYMFVV HFIIPLIVIF FCYGQLVFTV KEAAAQQQES

       250        260        270        280        290        300
ATTQKAEKEV TRMVIIMVIA FLICWLPYAG VAFYIFTHQG SDFGPIFMTI PAFFAKTSAV

       310        320        330        340
YNPVIYIMMN KQFRNCMVTT LCCGKNPLGD DEASTTVSKT ETSQVAPA
```

human β_2 adrenergic receptor			
UniProtKB	P07550		
length	413 amino acids		
N-terminus	1-34	**E2-loop**	175-196
TM I	35-58	**TM V**	197-220
C1-loop	59-71	**C3-loop**	221-274
TM II	72-95	**TM VI**	275-298
E1-loop	96-106	**E3-loop**	299-305
TM III	107-129	**TM VII**	306-329
C2-loop	130-150	**C-terminus**	330-413
TM IV	151-174		
sequence			

```
        10         20         30         40         50         60
MGQPGNGSAF LLAPNGSHAP DHDVTQERDE VWVVGMGIVM SLIVLAIVFG NVLVITAIAK

        70         80         90        100        110        120
FERLQTVTNY FITSLACADL VMGLAVVPFG AAHILMKMWT FGNFWCEFWT SIDVLCVTAS

       130        140        150        160        170        180
IETLCVIAVD RYFAITSPFK YQSLLTKNKA RVIILMVWIV SGLTSFLPIQ MHWYRATHQE

       190        200        210        220        230        240
AINCYANETC CDFFTNQAYA IASSIVSFYV PLVIMVFVYS RVFQEAKRQL QKIDKSEGRF

       250        260        270        280        290        300
HVQNLSQVEQ DGRTGHGLRR SSKFCLKEHK ALKTLGIIMG TFTLCWLPFF IVNIVHVIQD

       310        320        330        340        350        360
NLIRKEVYIL LNWIGYVNSG FNPLIYCRSP DFRIAFQELL CLRRSSLKAY GNGYSSNGNT

       370        380        390        400        410
GEQSGYHVEQ EKENKLLCED LPGTEDFVGH QGTVPSDNID SQGRNCSTND SLL
```

turkey β_1 adrenergic receptor			
UniProtKB	P07700		
length	483 amino acids		
N-terminus	1-38	**E2-loop**	180-205
TM I	39-67	**TM V**	206-231
C1-loop	68-76	**C3-loop**	232-285
TM II	77-103	**TM VI**	286-315
E1-loop	104-115	**E3-loop**	316-320
TM III	116-137	**TM VII**	321-343
C2-loop	138-155	**C-terminus**	344-483
TM IV	156-179		
sequence			

```
        10         20         30         40         50         60
MGDGWLPPDC GPHNRSGGGG ATAAPTGSRQ VSAELLSQQW EAGMSLLMAL VVLLIVAGNV

        70         80         90        100        110        120
LVIAAIGRTQ RLQTLTNLFI TSLACADLVM GLLVVPFGAT LVVRGTWLWG SFLCECWTSL

       130        140        150        160        170        180
DVLCVTASIE TLCVIAIDRY LAITSPFRYQ SLMTRARAKV IICTVWAISA LVSFLPIMMH

       190        200        210        220        230        240
WWRDEDPQAL KCYQDPGCCD FVTNRAYAIA SSIISFYIPL LIMIFVYLRV YREAKEQIRK

       250        260        270        280        290        300
IDRCEGRFYG SQEQPQPPPL PQHQPILGNG RASKRKTSRV MAMREHKALK TLGIIMGVFT

       310        320        330        340        350        360
LCWLPFFLVN IVNVFNRDLV PDWLFVFFNW LGYANSAFNP IIYCRSPDFR KAFKRLLCFP

       370        380        390        400        410        420
RKADRRLHAG GQPAPLPGGF ISTLGSPEHS PGGTWSDCNG GTRGGSESSL EERHSKTSRS

       430        440        450        460        470        480
ESKMEREKNI LATTRFYCTF LGNGDKAVFC TVLRIVKLFE DATCTCPHTH KLKMKWRFKQ

HQA
```

human D_3 receptor			
UniProtKB	P35462		
length	400 amino acids		
N-terminus	1-32	**E2-loop**	171-187
TM I	33-55	**TM V**	188-212
C1-loop	56-65	**C3-loop**	213-329
TM II	66-88	**TM VI**	330-351
E1-loop	89-104	**E3-loop**	352-366
TM III	105-126	**TM VII**	367-388
C2-loop	127-149	**C-terminus**	389-400
TM IV	150-170		

sequence (Isoform 1, P35462-1), canonical sequence

```
        10         20         30         40         50         60
MASLSQLSSH LNYTCGAENS TGASQARPHA YYALSYCALI LAIVFGNGLV CMAVLKERAL

        70         80         90        100        110        120
QTTTNYLVVS LAVADLLVAT LVMPWVVYLE VTGGVWNFSR ICCDVFVTLD VMMCTASILN

       130        140        150        160        170        180
LCAISIDRYT AVVMPVHYQH GTGQSSCRRV ALMITAVWVL AFAVSCPLLF GFNTTGDPTV

       190        200        210        220        230        240
CSISNPDFVI YSSVVSFYLP FGVTVLVYAR IYVVLKQRRR KRILTRQNSQ CNSVRPGFPQ

       250        260        270        280        290        300
QTLSPDPAHL ELKRYYSICQ DTALGGPGFQ ERGGELKREE KTRNSLSPTI APKLSLEVRK

       310        320        330        340        350        360
LSNGRLSTSL KLGPLQPRGV PLREKKATQM VAIVLGAFIV CWLPFFLTHV LNTHCQTCHV

       370        380        390        400
SPELYSATTW LGYVNSALNP VIYTTFNIEF RKAFLKILSC
```

sequence (Isoform 3, P35462-3), length: 367 amino acids

```
        10         20         30         40         50         60
MASLSQLSSH LNYTCGAENS TGASQARPHA YYALSYCALI LAIVFGNGLV CMAVLKERAL

        70         80         90        100        110        120
QTTTNYLVVS LAVADLLVAT LVMPWVVYLE VTGGVWNFSR ICCDVFVTLD VMMCTASILN

       130        140        150        160        170        180
LCAISIDRYT AVVMPVHYQH GTGQSSCRRV ALMITAVWVL AFAVSCPLLF GFNTTGDPTV

       190        200        210        220        230        240
CSISNPDFVI YSSVVSFYLP FGVTVLVYAR IYVVLKQRRR KRILTRQNSQ CNSVRPGFPQ

       250        260        270        280        290        300
QTLSPDPAHL ELKRYYSICQ DTALGGPGFQ ERGGELKREE KTRNSLMPLR EKKATQMVAI

       310        320        330        340        350        360
VLGAFIVCWL PFFLTHVLNT HCQTCHVSPE LYSATTWLGY VNSALNPVIY TTFNIEFRKA

FLKILSC
```

human H_1 receptor			
UniProtKB	P35367		
length	487 amino acids		
N-terminus	1-29	**E2-loop**	166-189
TM I	30-49	**TM V**	190-210
C1-loop	50-63	**C3-loop**	211-418
TM II	64-83	**TM VI**	419-438
E1-loop	84-101	**E3-loop**	439-450
TM III	102-123	**TM VII**	451-470
C2-loop	124-145	**C-terminus**	471-487
TM IV	146-165		

sequence

```
        10         20         30         40         50         60
MSLPNSSCLL EDKMCEGNKT TMASPQLMPL VVVLSTICLV TVGLNLLVLY AVRSERKLHT

        70         80         90        100        110        120
VGNLYIVSLS VADLIVGAVV MPMNILYLLM SKWSLGRPLC LFWLSMDYVA STASIFSVFI

       130        140        150        160        170        180
LCIDRYRSVQ QPLRYLKYRT KTRASATILG AWFLSFLWVI PILGWNHFMQ QTSVRREDKC

       190        200        210        220        230        240
ETDFYDVTWF KVMTAIINFY LPTLLMLWFY AKIYKAVRQH CQHRELINRS LPSFSEIKLR

       250        260        270        280        290        300
PENPKGDAKK PGKESPWEVL KRKPKDAGGG SVLKSPSQTP KEMKSPVVFS QEDDREVDKL

       310        320        330        340        350        360
YCFPLDIVHM QAAAEGSSRD YVAVNRSHGQ LKTDEQGLNT HGASEISEDQ MLGDSQSFSR

       370        380        390        400        410        420
TDSDTTTETA PGKGKLRSGS NTGLDYIKFT WKRLRSHSRQ YVSGLHMNRE RKAAKQLGFI

       430        440        450        460        470        480
MAAFILCWIP YFIFFMVIAF CKNCCNEHLH MFTIWLGYIN STLNPLIYPL CNENFKKTFK

RILHIRS
```

human CXCR4 receptor			
UniProtKB	P61073		
length	352 amino acids		
N-terminus	1-38	**E2-loop**	175-195
TM I	39-63	**TM V**	196-216
C1-loop	64-77	**C3-loop**	217-241
TM II	78-99	**TM VI**	242-261
E1-loop	100-110	**E3-loop**	262-282
TM III	111-130	**TM VII**	283-302
C2-loop	131-154	**C-terminus**	303-352
TM IV	155-174		

sequence (Isoform 1, P61073-1), canonical sequence

```
        10         20         30         40         50         60
MEGISIYTSD NYTEEMGSGD YDSMKEPCFR EENANFNKIF LPTIYSIIFL TGIVGNGLVI

        70         80         90        100        110        120
LVMGYQKKLR SMTDKYRLHL SVADLLFVIT LPFWAVDAVA NWYFGNFLCK AVHVIYTVNL

       130        140        150        160        170        180
YSSVLILAFI SLDRYLAIVH ATNSQRPRKL LAEKVVYVGV WIPALLLTIP DFIFANVSEA

       190        200        210        220        230        240
DDRYICDRFY PNDLWVVVFQ FQHIMVGLIL PGIVILSCYC IIISKLSHSK GHQKRKALKT

       250        260        270        280        290        300
TVILILAFFA CWLPYYIGIS IDSFILLEII KQGCEFENTV HKWISITEAL AFFHCCLNPI

       310        320        330        340        350
LYAFLGAKFK TSAQHALTSV SRGSSLKILS KGKRGGHSSV STESESSSFH SS
```

sequence (Isoform 2, P61073-2), length: 356 amino acids

```
        10         20         30         40         50         60
MSIPLPLLQI YTSDNYTEEM GSGDYDSMKE PCFREENANF NKIFLPTIYS IIFLTGIVGN

        70         80         90        100        110        120
GLVILVMGYQ KKLRSMTDKY RLHLSVADLL FVITLPFWAV DAVANWYFGN FLCKAVHVIY

       130        140        150        160        170        180
TVNLYSSVLI LAFISLDRYL AIVHATNSQR PRKLLAEKVV YVGVWIPALL LTIPDFIFAN

       190        200        210        220        230        240
VSEADDRYIC DRFYPNDLWV VVFQFQHIMV GLILPGIVIL SCYCIIISKL SHSKGHQKRK

       250        260        270        280        290        300
ALKTTVILIL AFFACWLPYY IGISIDSFIL LEIIKQGCEF ENTVHKWISI TEALAFFHCC

       310        320        330        340        350
LNPILYAFLG AKFKTSAQHA LTSVSRGSSL KILSKGKRGG HSSVSTESES SSFHSS
```

human A_{2A} receptor			
UniProtKB	P29274		
length	412 amino acids		
N-terminus	1-7	**E2-loop**	144-173
TM I	8-32	**TM V**	174-198
C1-loop	33-42	**C3-loop**	199-234
TM II	43-66	**TM VI**	235-258
E1-loop	67-77	**E3-loop**	259-266
TM III	78-100	**TM VII**	267-290
C2-loop	101-120	**C-terminus**	291-412
TM IV	121-143		
sequence			

```
        10         20         30         40         50         60
MPIMGSSVYI TVELAIAVLA ILGNVLVCWA VWLNSNLQNV TNYFVVSLAA ADIAVGVLAI

        70         80         90        100        110        120
PFAITISTGF CAACHGCLFI ACFVLVLTQS SIFSLLAIAI DRYIAIRIPL RYNGLVTGTR

       130        140        150        160        170        180
AKGIIAICWV LSFAIGLTPM LGWNNCGQPK EGKNHSQGCG EGQVACLFED VVPMNYMVYF

       190        200        210        220        230        240
NFFACVLVPL LLMLGVYLRI FLAARRQLKQ MESQPLPGER ARSTLQKEVH AAKSLAIIVG

       250        260        270        280        290        300
LFALCWLPLH IINCFTFFCP DCSHAPLWLM YLAIVLSHTN SVVNPFIYAY RIREFRQTFR

       310        320        330        340        350        360
KIIRSHVLRQ QEPFKAAGTS ARVLAAHGSD GEQVSLRLNG HPPGVWANGS APHPERRPNG

       370        380        390        400        410
YALGLVSGGS AQESQGNTGL PDVELLSHEL KGVCPEPPGL DDPLAQDGAG VS
```

Lysozyme (LYS_BPT4)	
UniProtKB	P00720
length	164 amino acids
organism	enterobacteria phage T4 (bacteriophage T4)
sequence	

```
        10         20         30         40         50         60
MNIFEMLRID EGLRLKIYKD TEGYYTIGIG HLLTKSPSLN AAKSELDKAI GRNCNGVITK

        70         80         90        100        110        120
DEAEKLFNQD VDAAVRGILR NAKLKPVYDS LDAVRRCALI NMVFQMGETG VAGFTNSLRM

       130        140        150        160
LQQKRWDEAA VNLAKSRWYN QTPNRAKRVI TTFRTGTWDA YKNL
```

bovine guanine-nucleotide binding protein G(t) subunit alpha-1 (GNAT1)	
UniProtKB	P04695
length	350 amino acids
sequence	

```
        10         20         30         40         50         60
MGAGASAEEK HSRELEKKLK EDAEKDARTV KLLLLGAGES GKSTIVKQMK IIHQDGYSLE

        70         80         90        100        110        120
ECLEFIAIIY GNTLQSILAI VRAMTTLNIQ YGDSARQDDA RKLMHMADTI EEGTMPKEMS

       130        140        150        160        170        180
DIIQRLWKDS GIQACFDRAS EYQLNDSAGY YLSDLERLVT PGYVPTEQDV LRSRVKTTGI

       190        200        210        220        230        240
IETQFSFKDL NFRMFDVGGQ RSERKKWIHC FEGVTCIIFI AALSAYDMVL VEDDEVNRMH

       250        260        270        280        290        300
ESLHLFNSIC NHRYFATTSI VLFLNKKDVF SEKIKKAHLS ICFPDYNGPN TYEDAGNYIK

       310        320        330        340        350
VQFLELNMRR DVKEIYSHMT CATDTQNVKF VFDAVTDIII KENLKDCGLF
```

bovine guanine-nucleotide binding protein G(s) subunit alpha isoform short (GNAS)

UniProtKB	P04896
length	394 amino acids

sequence (Isoform GNAS-1, Alpha-S2)

```
        10         20         30         40         50         60
MGCLGNSKTE DQRNEEKAQR EANKKIEKQL QKDKQVYRAT HRLLLLGAGE SGKSTIVKQM

        70         80         90        100        110        120
RILHVNGFNG EGGEEDPQAA RSNSDGEKAT KVQDIKNNLK EAIETIVAAM SNLVPPVELA

       130        140        150        160        170        180
NPENQFRVDY ILSVMNVPDF DFPPEFYEHA KALWEDEGVR ACYERSNEYQ LIDCAQYFLD

       190        200        210        220        230        240
KIDVIKQDDY VPSDQDLLRC RVLTSGIFET KFQVDKVNFH MFDVGGQRDE RRKWIQCFND

       250        260        270        280        290        300
VTAIIFVVAS SSYNMVIRED NQTNRLQEAL NLFKSIWNNR WLRTISVILF LNKQDLLAEK

       310        320        330        340        350        360
VLAGKSKIED YFPEFARYTT PEDATPEPGE DPRVTRAKYF IRDEFLRIST ASGDGRHYCY

       370        380        390
PHFTCAVDTE NIRRVFNDCR DIIQRMHLRQ YELL
```

sequence (Isoform GNAS-2, Alpha-S1), length: 380 amino acids

```
        10         20         30         40         50         60
MGCLGNSKTE DQRNEEKAQR EANKKIEKQL QKDKQVYRAT HRLLLLGAGE SGKSTIVKQM

        70         80         90        100        110        120
RILHVNGFNG DGEKATKVQD IKNNLKEAIE TIVAAMSNLV PPVELANPEN QFRVDYILSV

       130        140        150        160        170        180
MNVPDFDFPP EFYEHAKALW EDEGVRACYE RSNEYQLIDC AQYFLDKIDV IKQDDYVPSD

       190        200        210        220        230        240
QDLLRCRVLT SGIFETKFQV DKVNFHMFDV GGQRDERRKW IQCFNDVTAI IFVVASSSYN

       250        260        270        280        290        300
MVIREDNQTN RLQEALNLFK SIWNNRWLRT ISVILFLNKQ DLLAEKVLAG KSKIEDYFPE

       310        320        330        340        350        360
FARYTTPEDA TPEPGEDPRV TRAKYFIRDE FLRISTASGD GRHYCYPHFT CAVDTENIRR

       370        380
VFNDCRDIIQ RMHLRQYELL
```

rat guanine nucleotide-binding protein G(I)/G(S)/G(T) subunit beta-1 (GNB1)	
UniProtKB	P54311
length	340 amino acids
sequence	

```
        10         20         30         40         50         60
MSELDQLRQE AEQLKNQIRD ARKACADATL SQITNNIDPV GRIQMRTRRT LRGHLAKIYA

        70         80         90        100        110        120
MHWGTDSRLL VSASQDGKLI IWDSYTTNKV HAIPLRSSWV MTCAYAPSGN YVACGGLDNI

       130        140        150        160        170        180
CSIYNLKTRE GNVRVSRELA GHTGYLSCCR FLDDNQIVTS SGDTTCALWD IETGQQTTTF

       190        200        210        220        230        240
TGHTGDVMSL SLAPDTRLFV SGACDASAKL WDVREGMCRQ TFTGHESDIN AICFFPNGNA

       250        260        270        280        290        300
FATGSDDATC RLFDLRADQE LMTYSHDNII CGITSVSFSK SGRLLLAGYD DFNCNVWDAL

       310        320        330        340
KADRAGVLAG HDNRVSCLGV TDDGMAVATG SWDSFLKIWN
```

Bovine guanine nucleotide-binding protein G(I)/G(S)/G(O) subunit gamma-2 (GNG2)	
UniProtKB	P63212
length	71 amino acids
sequence	

```
        10         20         30         40         50         60
MASNNTASIA QARKLVEQLK MEANIDRIKV SKAAADLMAY CEAHAKEDPL LTPVPASENP

        70
FREKKFFCAI L
```

References

Abraham MH (1984) Thermodynamics of solution of homologous series of solutes in water. J Chem Soc Farad T 1 80:153–181

Alves ID, Salamon Z, Varga E, Yamamura HI, Tollin G, Hruby VJ (2003) Direct observation of G-protein binding to the human δ-opioid receptor using plasmon-waveguide resonance spectroscopy. J Biol Chem 278:48890–48897

Alves ID, Salago GFJ, Salamon Z, Brown MF, Tollin G, Hruby VJ (2005) Phosphatidylethanolamine enhances rhodopsin photoactivation and transducin binding in a solid supported lipid bilayer as determined using plasmon-waveguide resonance spectroscopy. Biophys J 88:198–210

Bae H, Cabrera-Vera TM, Depree KM, Graber SG, Hamm HE (1999) Two amino acids within the α4 helix of Gi1 mediate coupling with 5-hydroxytryptamine1B receptors. J Biol Chem 274:14963–14971

Ballesteros JA, Shi L, Javitch JA (2001) Structural mimicry in G protein-coupled receptors: implications of the high resolution crystal structure of rhodopsin for structure-function analysis of rhodopsin-like receptors. Mol Pharmacol 60:1–19

Barth P, Wallner B, Baker D (2009) Prediction of membrane protein structures with complex topologies using limited constraints. PNAS 106:1409–1414

Belohorcova K, Davis JH, Woolf TB, Roux B (1997) Structure and dynamics of an amphiphilic peptide in a lipid bilayer: a molecular dynamics study. Biophys J 73:3039–3055

Bokoch MP, Zou Y, Rasmussen SGF, Liu CW, Nygaard R, Rosenbaum DM, Fund JJ, Choi HJ, Thian FS, Kobilka TS, Puglisi JD, Weis WI, Pardo L, Prosser RS, Mueller L, Kobilka BK (2010) Ligand-specific regulation of the extracellular surface of a G-protein-coupled receptor. Nature 463:108–112

Bouzida D, Kumar S, Swendsen RH (1992) Efficient Monte Carlo methods for the computer simulation of biological molecules. Phys Rev A 45:8894–8901

Bräuner-Osborne H, Wellendorph P, Jensen AA (2007) Structure, pharmacology and therapeutic prospects of family C G protein coupled receptors. Curr Drug Targets 8:169–184

Brown N, Lewis RA (2006) Exploiting QSAR methods in lead optimization. Curr Opin Drug Discov Devel 9:419–424

Brunskole I, Strasser A, Seifert R, Buschauer A (2011) Role of the second and third extracellular loops of the histamine H_4 receptor in receptor activation. Naunyn Schmiedeberg's Arch Pharmacol 384:301–317

Cabani S, Gianni P, Mollica V, Lepori L (1981) Group contributions to the thermodynamic properties of non-ionic organic solutes in dilute aqueous solution. J Solution Chem 10:563–595

Cai K, Itoh Y, Khorana HG (2001) Mapping of contact sites in complex formation between transducin and light-activated rhodopsin by covalent crosslinking: use of a photoactivatable reagent. Proc Natl Acad Sci USA 98:4877–4882

Carloni P, Rothlisberger U, Parinello M (2002) The role and perspective of ab initio molecular dynamics in the study of biological systems. Acc Chem Res 35:455–464

A. Strasser, H.-J. Wittmann, *Modelling of GPCRs*,
DOI 10.1007/978-94-007-4596-4, © Springer Science+Business Media Dordrecht 2013

Chalmers DT, Behan DP (2002) The use of constitutively active GPCRs in drug discovery and functional genomics. Nature Rev Drug Discov 1:599–608

Cherezov V, Rosenbaum DM, Hanson MA, Rasmussen SGF, Thian FS, Kobilka TS, Choi HJ, Kuhn P, Weis WI, Kobilka BK, Stevens RS (2007) High-resolution crystal structure of an engineered human β_2-adrenergic G protein-coupled receptor. Science 318:1258–1265

Chien EY, Liu W, Zhao Q, Katritch V, Han GW, Hanson MA, Shi L, Newman AH, Javitch JA, Cherezov V, Stevens RC (2010) Structure of the human dopamine D_3 receptor in complex with D_2/D_3 selective antagonist. Science 330:1091–1095

Choe HW, Kim YJ, Park JH, Morizumi T, Pai EF, Krauß N, Hofmann KP, Scheerer P, Ernst OP (2011) Crystal structure of metarhosopsin II. Nature 471:651–655

Chou KC (2005) Coupling interaction thromboxane A_2 receptor and alpha-13 subunit of guanine nucleotide-binding protein. J Proteome Res 4:1681–1686

Christen M, van Gunsteren WF (2008) On searching in, sampling of and dynamically moving through conformational space of biomolecular systems: a review. J Comput Chem 29:157–166

Clark M, Cramer RD, van Opdenbosch N (1989) Validation of the general purpose tripos 5.2 force field. J Comput Chem 10:982–1012

Cornell WD, Cieplak B, Bayly CI, Gould IR, Merz KM, Ferguson DM, Spellmeyer DC, Fox T, Caldwell JW, Kollmann PA (1995) A second generation force field for the simulation of proteins, nucleic acids, and organic molecules. J Am Chem Soc 117:5179–5197

Costanzi S, Joshi BV, Maddileti S, Mamedova L, Gonzalez-Moa MJ, Marquez VE, Harden TK, Jacobson KA (2005) Human $P2Y_6$ receptor: molecular modeling leads to the rational design of a novel agonist based on a unique conformational preference. J Med Chem 48:8108–8111

Domanski J, Stansfeld P, Sansom MSP, Beckstein O (2010) Lipidbook: a public repository for force field parameters used in membrane simulations. J Membrane Biol 236:255–258

Dore AS, Robertson N, Errey JC, Ng I, Hollenstein K, Tehan B, Hurrell E, Bennett K, Congreve M, Magnani F, Tate CG, Weir M, Marshall FH (2011) Structure of the adenosine A_2A receptor in complex with ZM241385 and the xanthenes XAC and caffeine. Structure 19:1283–1293

Duan Y, Wu C, Chowdhury S, Lee MC, Xiong G, Zhang W, Yang R, Cieplak P, Luo R, Lee T, Caldwell J, Wang J, Kollman P (2003) A point-charge force field for molecular mechanics simulations of proteins based on condensed-phase quantum mechanical calculations. J Comput Chem 24:1999–2012

Dudek AZ, Arodz T, Galvez J (2006) Computational methods in developing quantitative structure-activity relationships (QSAR): a review. Comb Chem High Throughput Screen 9:213–228

Elefsinioti AL, Bagos PG, Spyropoulos IC, Hamodrakas SJ (2004) A database for G proteins and their interaction with GPCRs. Bioinformatics 5:208–215

Evers A, Klebe G (2004) Successful virtual screening for a submicromolar antagonist of the neurokinin-1 receptor based on a ligand-supported homology model. J Med Chem 47:5381–5392

Fanelli F, Menziani C, Scheer A, Cotecchia S, de Benedetti PG (1999) Theoretical study of the electrostatically driven step of receptor G-protein recognition. Proteins 37:145–156

Fanelli F, Menziani C, Scheer A, Cotecchia S, de Benedetti PG (1999) Theoretical study on receptor-G protein recognition: new insights into the mechanism of the α1b-adrenergic receptor activation. Int J Quant Chem 73:71–83

Filizola M, Wang SX, Weinstein H (2006) Dynamic models of G-protein coupled receptor dimmers: indications of asymmetry in the rhodopsin dimer from molecular dynamics simulations in a POPC bilayer. J Comput Aided Mol Des 20:405–416

Fleishmann SJ, Ben-Tal N (2006) Progress in structure prediction of a-helical membrane proteins. Curr Opin Struct Biol 16:496–504

Fredriksson R, Lagerström MC, Lundin LG, Schiöth HB (2003) The G-protein-coupled receptors in the human genome form five main families. Phylogenetic analysis, paralogon groups, and fingerprints. Mol Pharmacol 63:1256–1272

Frenkel D, Smit B (2002) Understanding molecular simulation – from algorithms to applications. Academic Press, San Diego

Gales C, van Durm JJJ, Schaak S, Pontier S, Percheranchier Y, Audet M, Paris H, Bouvier M (2006) Probing the activation-promoted structural rearrangements in preassembled receptor-G-protein complexes. Nat Struct Mol Biol 13:778–786

Ganjiwale AD, Rao GS, Cowsik SM (2011) Molecular modeling of neurokinin B and tachykinin NK3 receptor complex. J Chem Inf Model 51:2932–2938

Gascon JA, Leung SSF, Batista ER, Batista VS (2006) A self-consistent space-domain decomposition method for QM/MM computations of protein electrostatic potentials. J Chem Theory Comput 2:175–186

Gedeck P, Lewis RA (2008) Exploiting QSAR models in lead optimization. Curr Opin Drug Discov Devel 11:569–575

Gether U, Kobilka BK (1998) G protein coupled receptors. II. Mechanism of agonist activation. J Biol Chem 273:17979–17982

Goetz A, Lanig H, Gmeiner P, Clark T (2011) Molecular dynamics simulations of the effect of the G-protein and diffusible ligands on the β_2-adrenergic receptor. J Mol Biol 414:611–623

Greasley PJ, Fanelli F, Scheer A, Abuin L, Nenniger-Tosato M, de Benedetti PG, Cotecchia S (2001) Mutational and computational analysis of the α_1b-adrenergic receptor: involvement of basic and hydrophobic residues in receptor activation and G protein coupling. J Biol Chem 276:46485–46494

Grishina G, Bertlot CH (2000) A surface-exposed region of $G_s\alpha$ in which substitutions decrease receptor-mediated activation and increase receptor affinity. Mol Pharmacol 57:1081–1092

van Gunsteren WF, Berendsen HJC (1987) Thermodynamic cycle integration by computer simulation as a tool for obtaining free energy differences in molecular chemistry. J Comput Aided Mol Des 1:171–176

van Gunsteren WF, Berendsen HJC (1990) Computer simulation of molecular dynamics: methology, applications, and perspectives in chemistry. Angew Chem Int Ed Engl 29:992–1023

van Gunsteren WF, Billeter SR, Eising AA, Hünenberger PH, Krüger P, Mark AE, Scott WRP, Tironi IG (1996) Biomolecular simulation: the GROMOS96 manual an user guide. Hochschulverlag AG an der ETH, Zürich

Halgren TA (1996) Merck molecular force field. I. Basis, form, scope, parameterization, and performance of MMFF94. J Comput Chem 17:490–519

Hanson MA, Cherezov V, Griffith MT, Roth CB, Jaakola VP, Chien YET, Velasquez J, Kuhn P, Stevens RC (2008) A specific cholesterol binding site is established by the 2.8 Å structure of the human β_2-adrenergic receptor. Structure 16:897–905

Harmar AJ (2001) Family-B G-protein-coupled receptors. Genome Biology 2:REVIEWS2012

Henin J, Maigret B, Tarek M, Escrieut C, Fourmy D, Chipot C (2006) Probing a model of a GPCR/ligand complex in an explicit membrane environment: the human cholecystokinin-1 receptor. Biophys J 90:1232–1240

Herrmann R, Heck M, Henklein P, Henklein P, Kleuss C, Hofmann KP, Ernst OP (2004) Sequence of interactions in receptor-G protein coupling. J Biol Chem 279:24283–24290

Herrmann R, Heck M, Henklein P, Hofmann KP, Ernst OP (2006) Signal transfer from GPCRs to G proteins: role of the Gα N-terminal region in rhodopsin-transducin coupling. J Biol Chem 281:30234–30241

Hoare SRJ (2005) Mechanisms of peptide and nonpeptide ligand binding to class B G-protein-coupled receptors. Drug Discov Today 10:417–427

Igel P, Geyer R, Strasser A, Dove S, Seifert R, Buschauer A (2009) Synthesis and structure-activity relationships of cyanoguanidine-type and structurally related histamine H_4 receptor agonists. J Med Chem 52:6297–6313

Isralewitz B, Izrailev S, Schulten K (1997) Binding pathway of retinal to bacterio-opsin: a prediction by molecular dynamics simulation. Biophys J 73:2972–2979

IUPHAR (2000) Committee on receptor nomenclature and drug classification, The IUPHAR compendium of receptor characterization and classification, 2nd edn. IUPHAR Media, London

Ivanov AA, Baskin II, Palyulin VA, Piccagli L, Baraldi PG, Zefirov NS (2005) Molecular modelling and molecular dynamics simulation of the human A_2B adenosine receptor. The study of the possible binding modes of the A_2B receptor antagonists. J Med Chem 48:6813–6820

Jaakola VP, Griffith MT, Hanson MA, Cherezov V, Chien YET, Lane JR, Ijzerman AP, Stevens RC (2008) The 2.6 angstrom crystal structure of a human A_2A adenosine receptor bound to an antagonist. Science 322:1211–1217

Jacoby E, Bouhelal R, Gerspacher M, Seuwen K (2006) The 7TM G-protein-coupled receptor target family. ChemMedChem 1:760–782

Jensen F (1999) Introduction to computational chemistry. Wiley, Chichester

Jongejan A, Bruysters M, Ballesteros JA, Haaksma E, Bakker RA, Pardo L, Leurs R (2005) Linking agonist binding to histamine H_1 receptor activation. Nat Chem Biol 1:98–103

Jongejan A, Lim HD, Smits RA, de Esch IJP, Haaksma E, Leurs R (2008) Delineation of agonist binding to the human histamine H_4 receptor using mutational analysis, homology modeling, and ab initio calculations. J Chem Inf Model 48:1455–1463

Jorgensen WL, Maxwell DS, Tirado-Rives J (1996) Development and testing of the OPLS all-atom force field on conformational energetics and properties of organic liquids. J Am Chem Soc 118:11225–11236

Kenakin T (1997) Pharmacologic analysis of drug receptor interaction. Lippincott Williams and Wilkins, New York

Khelashvili G, Dorff K, Shan J, Camacho-Artacho M, Skrabanek L, Vroling B, Bouvier M, Devi LA, George SR, Javitch JA, Lohse MJ, Milligan G, Neubig RR, Palczewski K, Parmentier M, Pin JP, Vriend G, Campagne F, Filizola M (2010) GPCR-OKB: the G protein coupled receptor oligomer knowledge base. Bioinformatics 26:1804–1805

Klotz IM, Rosenberg RM (2008) Chemical thermodynamics – basic concepts and methods, 7th edn. Wiley, Hoboken

Kobilka BK, Deupi X (2007) Conformational complexity of G-protein-coupled receptors. TRENDS Pharmacol Sci 28:397–406

Kondepudi P, Prigognie I (1998) Modern thermodynamics from heat engines to dissipative structures. Wiley, New York

Kosztin D, Izrailev S, Schulten K (1999) Unbinding of retinoic acid from its receptor studied by steered molecular dynamics. Biophys J 76:188–197

Kristiansen K (2004) Molecular mechanisms of ligand binding, signalling, and regulation within the superfamily of G-protein coupled receptors: molecular modelling and mutagenesis approaches to receptor structure and function. Pharmacol Ther 103:21–80

Kubinyi H (2011) 3D QSAR in drug design, Springer, Berlin

Kukol A (ed) (2010) Molecular modeling of proteins. Humana Press, New York

Leavitt S, Freire E (2001) Direct measurement of protein binding energetics by isothermal titration calorimetry. Curr Opin Struct Biol 11:560–566

Lebon G, Warne T, Edwards PC, Bennett K, Langmead CJ, Leslie AGW, Tate CG (2011) Agonist-bound adenosine A_2A receptor structures reveal common features of GPCR activation. Nature 474:521–525

Li J, Edwards PC, Burghammer M, Villa C, Schertler GFX (2004) Structure of bovine rhodopsin in a trigonal crystal form. J Mol Biol 343:1409–1438

Lim HD, Jongejan A, Bakker RA, Haaksma E, de Esch IJP, Leurs R (2008) Phenylalanine 169 in the second extracellular loop of the human histamine H_4 receptor is responsible for the difference in agonist binding between human and mouse H_4 receptors. J Pharmacol Exp Ther 327:88–96

Lipkowitz KB, Boyd DB (2007) Semiempirical molecular orbital methods. Reviews Comp Chem 1:45–81

McKerell AD, Bashford D, Bellott M, Dunbrack RL, Evanseck JD, Field MJ, Fischer S, Gao J, Guo H, Ha S, McCarthy JD, Kuchnir L, Kuczera K, Lau FTK, Mattos C, Michnick S, Ngo T, Nguyen DT, Prodhom B, Reiher WE, Roux B, Schlenkrich M, Smith JC, Stote R, Straub J, Watanabe M, Wiorkiewicz-Kuczera K, Yin D, Karplus M (1998) All-atom empirical potential for molecular modeling and dynamics studies of proteins. J Phys Chem B 102:3586–3616

Mackerell AD (2004) Empirical force fields for biological macromolecules: overview and issues. J Comput Chem 25:1584–1604

Mehler EL, Hassan SA, Kortagere S, Weinstein H (2006) Ab initio computational modeling of loops in G-protein-coupled receptors: lessons from the crystal structure of rhodopsin. Prot Struct Funct Bioinfor 64:673–690

Menikarachchi LC, Gascon JA (2008) Optimization of cutting schemes for the evaluation of molecular electrostatic potentials in proteins via Moving-Domain QM/MM. J Mol Model 14:1–9

Metropolis N (1987) The beginning of the Monte Carlo method. Los Alamos Sci 12:125–130

Monard G, Merz KM (1999) Combined quantum mechanical/molecular mechanical methodologies applied to biomolecular systems. Acc Chem Res 32:904–911

Moukhametzianov R, Warne T, Edwards PC, Serrano-Vega MJ, Leslie AGW, Tate CG, Schertler GFX (2011) Two distinct conformations of helix 6 observed in antagonist-bound structures of a β_1-adrenergic receptor. Proc Natl Acad Sci USA 108:8228–8232

Mustafi D, Palczewski K (2009) Topology of class A G protein-coupled receptors: insights gained from crystal structures of rhodopsins, adrenergic and adenosine receptors. Mol Pharmacol 75:1–12

Oldham WM, Hamm HE (2006) Structural basis of function in heterotrimeric G proteins. Q Rev Biophys 39:117–166

Oldham WM, Hamm HE (2008) Heterotrimeric G protein activation by G-protein-coupled receptors. Nat Rev Mol Cell Biol 9:60–71

Oliveira L, Paiva PB, Paiva ACM, Vriend G (2003) Sequence analysis reveals how G protein-coupled receptors transduce the signal to the G protein. Proteins 52:553–560

Oostenbrink C, Villa A, Mark AE, van Gunsteren WF (2004) A biomolecular force field based on the free enthalpy of hydration and solvation: the GROMOS force-field parameter sets 53A5 and 53A6. J Comput Chem 25:1656–1676

Overington JP, Bissan AL, Hopkins L (2006) How many drug targets are there? Nat Rev Drug Discov 5:993–996

Palczewski K, Kumasaka T, Hori T, Behnke CA, Motoshima H, Fox BA, LeTrong I, Teller DC, Okada T, Stenkamp RE, Yamamoto M, Miyano M (2000) Crystal structure of rhodopsin: a G protein-coupled receptor. Science 289:739–745

Pardo D, Deupi X, Dölker N, Lopez-Rodriguez ML, Campillo M (2007) The role of internal water molecules in the structure and function of the rhodopsin familiy of G protein-coupled receptors. Chem Bio Chem 8:19–24

Park JH, Scheerer P, Hofmann KP, Choe HW, Ernst OP (2008) Crystal structure of the ligand-free G protein-coupled receptor opsin. Nature 454:183–187

Pei Y, Mercier RW, Anday JK, Thakur GA, Zvonok AM, Hurst D, Reggio PH, Janero DR, Makriyannis A (2008) Ligand-binding architecture of human CB2 cannabinoid receptor: evidence for receptor subtype-specific binding motif and modelling GPCR activation. Chem Biol 15:1207–1219

Pierce KL, Premont RT, Lefkowitz RJ (2002) Seven transmembrane receptors. Nat Rev Mol Cell Biol 3:639–650

Preuss H, Ghorai P, Kraus A, Dove S, Buschauer A, Seifert R (2007) Point mutations in the second extracellular loop of the histamine H_2 receptor do not affect the species-selective activity of guanidine-type agonists. Naunyn Schmiedeberg's Arch Pharmacol 376:253–264

Raimondi F, Seeber M, de Benedetti PG, Fanelli F (2008) Mechanisms of the inter and intramolecular communication in GPCRs and G proteins. J Am Chem Soc 130:4310–4325

Rasmussen SGF, Choi HJ, Rosenbaum DM, Kobilka TS, Thian FS, Edwards PC, Burghammer M, Ratnala VRP, Sanishvili R, Fischetti RF, Schertler GFX, Weis WI, Kobilka BK (2007) Crystal structure of the human β_2 adrenergic G protein-coupled receptor. Nature 450:383–387

Rasmussen SG, Choi HJ, Fung JJ, Pardon E, Casarosa P, Chae PS, DeVree BT, Rosenbaum DM, Thian FS, Kobilka TS, Schnapp A, Konetzi J, Sunahara RK, Gellman SH, Pautsch A, Steyaert J, Weis WI, Kobilka BK (2011) Structure of a nanobody-stabilized active state of the β_2 -adrenoceptor. Nature 469:175–180

Rasmussen SGF, DeVree BT, Zou Y, Kruse AC, Chung KY, Kobilka TS, Thian FS, Chae PS, Pardon E, Calinski D, Mathiesen JM, Shah STA, Lyons JA, Caffrey M, Gellman SH, Steyaert J, Skiniotis G, Weis WI, Sunahara RK, Kobilka BK (2011) Crystal structure of the β_2 adrenergic receptor-Gs protein complex. Nature 477:549–555

Rosenbaum DM, Zhang C, Lyons JA, Holl R, Arago D, Arlow DH, Rasmussen SGF, Choi HJ, DeVree BT, Sunahara RK, Chae PS, Gellman SH, Dror RO, Shaw DE, Weis WI, Caffrey M, Gmeiner P, Kobilka BK (2011) Structure and function of an irreversible agonist-β_2 adrenoceptor complex. Nature 459:236–240

Scheerer P, Park JH, Hildebrand PW, Kim YJ, Krauß N, Choe HW, Hofmann KP, Ernst OP (2008) Crystal structure of opsin in its G-protein-interacting conformation. Nature 455:497–502

Schuettelkopf AW, van Aalten DMF (2004) PRODRG – a tool for high-throughput crystallography of protein-ligand complexes. Acta Crystallogr D60:1355–1363

Scior T, Medina-Franco JL, Do QT, Martinez-Mayorga K, Yunes Rojas JA, Bernard P (2009) How to recognize and workaround pitfalls in QSAR studies: a critical review. Curr Med Chem 16:4297–4313

Scott WRP, Huenenberger PH, Tironi IG, Mark AE, Billeter SR, Fennen J, Torda AE, Huber T, Krueger P, van Gunsteren WF (1999) The GROMOS biomolecular simulation program package. J Phys Chem A 103:3596–3607

Shimamura T, Shiroishi M, Weyand S, Tsuhimoto H, Winter G, Katritch V, Abagyan R, Cherezov V, Liu W, Han GW, Kobayashi T, Stevens RC, Iwata S (2011) Structure of the human histamine H_1 receptor complex with doxepin. Nature 475:65–70

Silberg RJ, Alberty RA, Bawendi MG (2005) Physical chemistry, 4th edn. Wiley, New York

Silva ME, Heim R, Strasser A, Elz S, Dove S (2011) Theoretical studies on the interaction of partial agonists with the 5-HT2A receptor. J Comput Aided Mol Des 25:51–66

Simpson LM, Taddese B, Wall ID, Reynolds CA (2010) Bioinformatics and molecular modelling approaches to GPCR oligomerization. Curr Opin Pharmacol 10:30–37

Skrabanek L, Murcia M, Bouvier M, Devi L, George SR, Lohse MJ, Milligan G, Neubig R, Palczewski K, Parmentier M, Pin JP, Vriend G, Javitch JA, Campagne F, Filizol M (2007) Requirements and ontology for a G protein-coupled receptor oligomerization knowledge base. BMC Bioinformatics 8:177

Stewart JJP (1989) Optimization of parameters for semiempirical methods II. Applications. J Comput Chem 10:221–264

Stewart JJP (2004) Optimization of parameters for semiempirical methods IV: extension of MNDO, AM1, and PM3 to more main group elements. J Mol Model 10:155–164

Strasser A, Wittmann HJ (2007a) LigPath: a module for predictive calculation of a ligands pathway into a receptor – application to the gpH_1-receptor. J Mol Model 13:209–218

Strasser A, Wittmann HJ (2007b) Analysis of the activation mechanism of the guinea-pig histamine H_1 receptor. J Comput Aided Mol Des 21:499–509

Strasser A, Wittmann HJ (2010) In silico analysis of the histaprodifen induced pathway of the guinea-pig histamine H_1 receptor. J Comput Aided Mol Des 24:759–769

Strasser A, Wittmann HJ (2010a) 3D-QSAR CoMFA study to predict orientation of suprahistaprodifens and phenoprodifens in the binding-pocket of four histamine H_1-receptor species. Mol Inf 29:333–341

Strasser A, Wittmann HJ (2010b) Distinct interactions between the human adrenergic β_2 receptor and $G\alpha_s$ – an in silico study. J Mol Model 16:1307–1318

Strasser A, Wittmann HJ (in press) hβ_2 R-$G\alpha_s$-complex: prediction versus crystal structure – how valuable are predictions based on molecular modelling studies? J Mol Model, in press

Suwa M, Sugiharar M, Ono Y (2011) Functional and structural overview of G protein coupled receptors comprehensively obtained from genome sequences. Pharmaceutical 4:652–664

Taylor MS, Fung HK, Rajgaria R, Filizola M, Weinstein H, Floudas CA (2008) Mutations affecting the oligomerization interface of G-protein-couple receptors revealed by a novel de novo protein design framework. Biophys J 94:2470–2481

Teller DC, Okada T, Behnke CA, Palczewski K, Stenkamp R (2001) Advance in determination of a high-resolution three-dimensional structure of rhodopsin, a model of G protein-coupled receptors (GPCRs). Biochemistry 40:7761–7772

Theodoropoulou M, Bagos PG, Spyropoulos IC, Hamodrakas SJ (2008) gpDB: a database of GPCRs, G-proteins, effectors and their interactions. Bioinformatics 24:1471–1472

Tolkovsky AM, Levitzki A (1978) Mode of coupling between the b-adrenergic receptor and adenylate cyclase in turkey erythrocytes. Biochemistry 17:3795–3810

Torres FE, Recht MI, Coyle JE, Bruce RH, Williams G (2010) Higher throughput calorimetry: opportunities, approaches and challenges. Curr Opin Struct Biol 20:598–605

Van Der Spoel D, Lindahl E, Hess B, van Buuren AR, Apol E, Meulenhoff PJ, Tieleman DP, Sijbers ALTM, Feenstra KA, van Drunen R, Berendsen HJC (2005) Gromacs User Manual version 4.0, http://www.gromacs.org

Vauquelin G, von Mentzer B (2007) G Protein-coupled receptors. Wiley-Blackwell, Wiley

Villa A, Mark AE (2002) Calculation of the free energy of solvation for neutral analogs of amino acid side chains. J Comput Chem 23:548–553

Vriend G (1990) WHAT IF: a molecular modeling and drug design program. J Mol Graph 8:52–56

Wacker D, Genalti G, Brown MA, Katritch V, Abagyan R, Cherezov V, Stevens RC (2010) Conserved binding mode of human 2 adrenergic receptor inverse agonists and antagonist revealed by X-ray crystallography. J Am Chem Soc 132:11443–11445

Wagner E, Wittmann HJ, Elz S, Strasser A (2011) Mepyramine-JNJ7777120-hybrid compounds show high affinity to hH_1R, but low affinity to hH_4R. Bioorg Med Chem Lett 21:6274–6280

Warne A, Serrano-Vega MJ, Baker JG, Moukhametzianov R, Edward PC, Henderson R, Leslie AGW, Tate CG, Schertler GFX (2008) Structure of the β_1-adrenergic G protein-coupled receptor. Nature 454:486–491

Warne A, Moukhametzianov R, Baker JG, Nehme R, Edwards PC, Leslie AGW, Schertler GFX, Tate CG (2011) The structural basis for agonist and partial agonist action on a β_1-adrenergic receptor. Nature 469:241–244

Weiland GA, Minneman KP, Molinoff PB (1979) Fundamental difference between the molecular interactions of agonists and antagonists with the β-adrenergic receptor. Nature 281:114–117

Wittmann HJ, Seifert R, Strasser A (2009) Contribution of binding enthalpy and entropy to affinity of antagonist and agonist binding at human and guinea-pig histamine H_1-receptor. Mol Pharmacol 76:25–37

Wittmann HJ, Seifert R, Strasser A (2011) N^{α} -methylated phenylhistamines exhibit affinity to the hH_4R – a pharmacological and molecular modelling study. Naunyn Schmiedeberg's Arch Pharmacol 384:287–299

Wise A, Gearing K, Rees S (2002) Target validation of G-protein coupled receptors. Drug Discov Today 7:235–246

Woolf TB, Roux B (1996) Structure, energetics, and dynamics of lipid-protein interactions – a molecular-dynamics study of the gramicidin-A channel in a DMPC bilayer. Prot Struc Func Genet 24:92–114

Wu B, Chien YET, Mol CD, Fenalti G, Liu W, Katritch V, Abagyan R, Brooun A, Wells P, Bi FC, Hamel DJ, Kuhn P, Handel TM, Cherezov V, Stevens RC (2010) Structures of the CXCR4 chemokine GPCR with small molecules and cyclic peptide antagonists. Science 330:1066–1071

Xu F, Wu H, Katritch V, Han GW, Jacobson KA, Gao ZG, Cherezov V, Stevens RC (2011) Agonist bound structure of the human adenosine A_{2A} receptor. Science 332:322–327

Yarov-Yarovoy V, Schonbrun J, Baker D (2006) Multipass membrane structure prediction using Rosetta. Prot Struct Funct Bioinfor 62:1010–1025

Zaki MJ, Bystroff C (2010) Protein structure prediction, 2nd edn. Human Press Inc., New York

Zhang Y, DeVries ME, Skolnick J (2006) Structure modeling of all identified G protein-coupled receptors in the human genome. PLOS Comput Biol 2:88–99

Zhang Y (2008) Progress and challenges in protein structure prediction. Curr Opin Struct Biol 18:342–348

Index

A. Strasser, H.-J. Wittmann, *Modelling of GPCRs*,
DOI 10.1007/978-94-007-4596-4, © Springer Science+Business Media Dordrecht 2013

MIX
Papier aus verantwortungsvollen Quellen
Paper from responsible sources
FSC® C105338

If you have any concerns about our products,
you can contact us on
ProductSafety@springernature.com

In case Publisher is established outside the EU,
the EU authorized representative is:
Springer Nature Customer Service Center GmbH
Europaplatz 3, 69115 Heidelberg, Germany

Printed by Libri Plureos GmbH
in Hamburg, Germany